Abhijit Bandyopadhyay, Poulomi Dasgupta, Sayan Basak
Mechanisms of Self-Healing Polymers

Also of interest

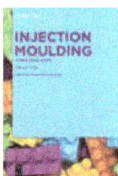

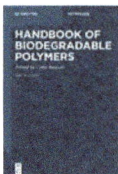

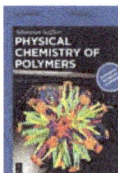

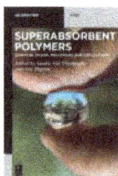

Abhijit Bandyopadhyay, Poulomi Dasgupta, Sayan Basak

Mechanisms of Self-Healing Polymers

Thermoplastics, Elastomers, TPEs and TPVs

DE GRUYTER

Authors
Prof. Abhijit Bandyopadhyay
Department of Polymer Science and Technology
University of Calcutta
92 A.P.C. Road
Kolkata 700009, West Bengal
India
abpoly@caluniv.ac.in

Poulomi Dasgupta
Department of Polymer Science and Technology
University of Calcutta
92 A.P.C. Road
Kolkata 700009, West Bengal
India
poulamidasgupta1992@gmail.com

Dr. Sayan Basak
Department of Polymer Science and Technology
University of Calcutta
92 A.P.C. Road
Kolkata 700009, West Bengal
India
sayancupst@gmail.com

ISBN 978-3-11-158309-9
e-ISBN (PDF) 978-3-11-158371-6
e-ISBN (EPUB) 978-3-11-158420-1

Library of Congress Control Number: 2025942169

Bibliographic information published by the Deutsche Nationalbibliothek
The Deutsche Nationalbibliothek lists this publication in the Deutsche Nationalbibliografie;
detailed bibliographic data are available on the internet at http://dnb.dnb.de.

© 2025 Walter de Gruyter GmbH, Berlin/Boston, Genthiner Straße 13, 10785 Berlin
Cover image: Christoph Burgstedt/iStock/Getty Images Plus
Typesetting: Integra Software Services Pvt. Ltd.

www.degruyterbrill.com
Questions about General Product Safety Regulation:
productsafety@degruyterbrill.com

Contents

List of figures

https://doi.org/10.1515/9783111583716-203

List of tables

https://doi.org/10.1515/9783111583716-204

Chapter 1
Introduction

The ability to regenerate is an oblivion aspect of research and technology. Technologists are being inundated by the discovery of self-healing material which enables them to prepare composites that can restore the structural and functional properties after the damage [1]. "Self-healing" was the term that started evolving in the late 90s mainly in the field of biological science and its prospective applications [2]. Back then the self-healing property was dependent on the amount of healing agents embedded in the microcapsule on the microtubule [3]. But several complications came up with the synthesis and delivery process such as altered target location, the optimum quantity of self-healing agents, and non-proctored pathway. As technology evolved, these certain limitations were undermined by modern dynamic chemistry, which is circumspect on the dynamic covalent chemistry, for instance, the imine bonds, disulfide bonds, acylhydrazone bonds, π–π stacking, hydrophobic and non-covalent interaction such as host–guest interaction [4, 5].

Graphene, which is primarily a two-dimensional planer monolayer of sp^2 carbon atoms, has overwhelmed the researchers with its enhanced mechanical, electrical, and thermal properties due to its large specific surface area [6]. As we know, polymeric materials generally form the self-healing lease materials due to their long-range elasticity and high recovery rate; graphene or graphene derivatives are widely amalgamated with these matrices inflating the overall property performance in content to the conductive device, coating, and pharmaceutical fields [7]. The incorporation of graphene or its derivative into the polymer matrix will endow them with the ability to repair themselves after damage and extend their overall service life [8].

1.1 Growth and development of self-healing materials and polymers

As the market for the polymer material increases, the end-user and the applications of these polymer composites also increase. However, these materials are prone to damage by an aggregation of external factors such as mechanical induction, chemical reaction, thermal exposures, and UV radiation [9]. On the other hand, there are only a few numbers of methods available to these structural composites if damaged, for repair and henceforth extending their functional lifetime. An ideal repair method is one that not only can be carried out quickly and directly on the cracked surface but also eliminates the question of replacement [10]. That is why it is very important for researchers to investigate the type of damage caused. For instance, a crack in a matrix may be repaired by sealing the crack (preferably with a resin), while a crack in fiber

https://doi.org/10.1515/9783111583716-001

will need a new fiber replacement or a fabric patch to achieve a similar level of mechanical properties [10, 11].

To trace back the history keeping the self-healing property as the backdrop, the ancient Romans used a modified form of lime mortar that had the ability to repair cracks autonomously [12]. In the year 2014, Marie Jackson, an American geologist, and her team recreated and investigated further the similar form of the mortar used in Trajan's Market and other famous Roman structures such as the Colosseum and the Pantheon. The studies revealed that the Romans blended a specific type of volcanic ash called the "Pozzolane Rosse" derived from the Alban Hills volcano with quicklime and water [13]. The Romans used the same blend to bind together the decametric-sized tuffs (an aggregate of volcanic rocks). As the material started to mature, with the induction of the Pozzolanic activity, the lime reacted with the other chemicals in the amalgamated mixture and in turn was replaced by fine crystals of calcium aluminosilicates (stratlingite). These crystals of the so-called stratlingite grew at the interfacial zones of the cementitious matrix in the material, where the cracks usually tend to develop. As the theory describes, this nucleating crystal formation would hold together the mortar and the coarse materials, giving rise to a new composite-like structure that had the ability to last for more than 1,900 years [14, 15].

After the Industrial Revolution, one of the popular methods to heal the fractured surfaces was "hot plate" welding, where the cracked surfaces were brought into contact above the glass transition temperature which allowed the process of interdiffusion to be carried out in order to repair the fractured surface restoring the strength of the material. The process faced a serious consequence, which involved the formation of the weld line, which in turn remained as the weakest point in the entire surface of the matrix [16]. For laminate composites, resin injection was often used to heal the damage by the process of delamination. Once again, the process possessed enough reasons to be obsolete because the crack at times was not accessible for injection [17]. None of these techniques provided an ideal solution to the healing problem that had presented. Although the above healing mechanisms do heal the fractured parts, they are temporary. On an additional note, the methods require monitoring of the fracture and manual controlling to execute the repairing strategies, thus elevating the overall cost of the material.

These drawbacks forced the researchers to investigate alternative and efficient novel healing techniques. In the year 1981, the word "self-healing polymeric material" was proposed to address the microcracks being developed within the polymer composites to improve the working life and safety of the material [18]. Further investigations on these smart materials in the year 1993 by Dry and Sattos promoted further insights into these self-healing polymeric materials [19]. White et al. in the year 2001 fabricated a novel material involving a microencapsulated healing agent that was released upon crack development. Furthermore, the US Air force and the European Space agency demonstrated several interesting pieces of research in the context of the self-healing materials [20, 21].

The twenty-first century has witnessed several advanced developments in the fields of self-healing polymers and allied materials. To add on, most of these recent advancements in the new class of smart materials were inspired by the biological systems since almost every material in nature has the inherent property of self-healing autonomously [22].

The number of publications revolving on the self-healing material has experienced a rapid surge in recent years, which proves the importance of these smart materials and the need for a self-healing strategy in this recent decade. Figures 1.1a and 1.1b demonstrate the number of publications as a function of time (time).

1.2 Commercial importance and the global self-healing material market trend

Damages to a polymer at a microstructural level, which are not easily detectable, cause a huge problem to the system in terms of its longevity and mechanical properties. The primary reason behind it is that the cracks cannot be manually repaired, and control becomes tortuous. Self-healing polymers inherit this property of autonomously healing the microcracks in presence of an external stimulus (sometimes without stimuli) to prevent further propagation of this damage, increasing the shelf life of the material. The Grand View Research (USA) had carried out a survey, which reported the market value of the self-healing material to be $120.4 [23].

Latest technologies such as the use of hollow glass fiber, optical fiber, or microcapsules have enabled science to create specialty polymer composites both using intrinsic and extrinsic types. Lowering the cost of these super-specialty materials has paved the way for the increase in the use of these materials.

Fig 1.2 shows an estimated growth of the increase of these smart materials proving their viability in the multidimensional field of applications. The predicted 78.02% increase from the year 2014 to the year 2025 has demonstrated the constant research and developmental works, which every laboratory is taking into consideration with its cutting-edge technological tools. The customer base is increased year after year by capturing the muted markets with the implementation of such sophisticated and smart materials [24].

In the context of self-healing, microcapsules have attracted several scientists for their property of in situ self-healing mechanism. Microcapsules generally release certain agents when ruptured causing an in situ polymerization that fixes the mechanical dent produced [25]. Microcapsules are made to be reinforced enough to survive the manufacturing process and consumption of the substances in their desired application.

These materials lessen the frequency of the maintenance costs of the material and bring the number of renovations down, which in turn is believed to be the driving factor for the growth of the market. Moreover, the global productions of self-

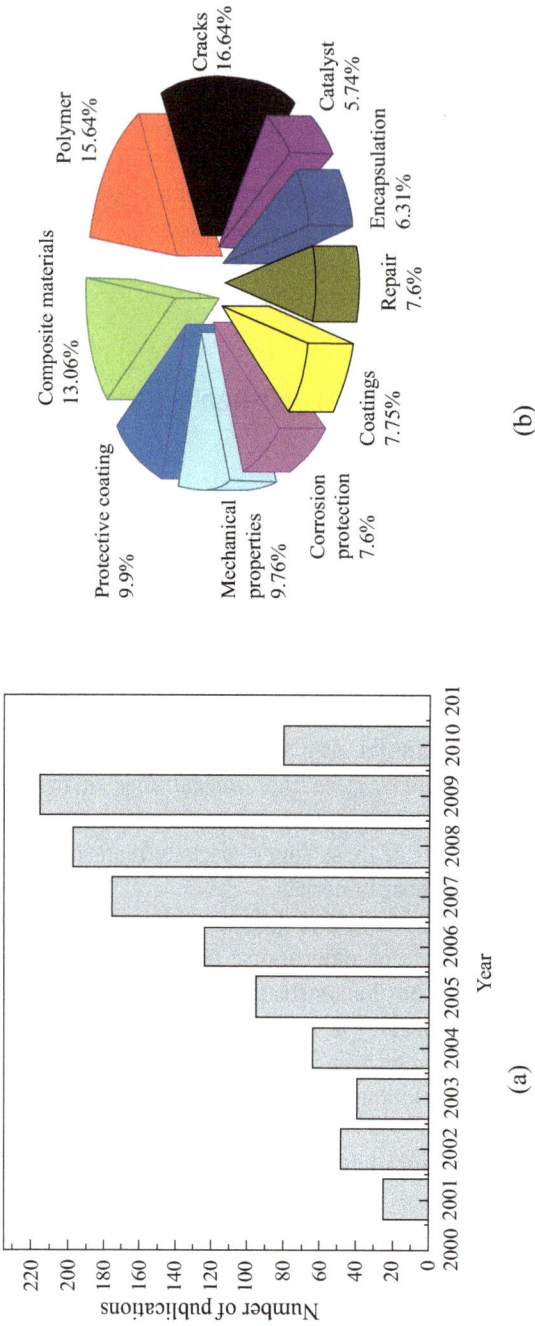

(a)

(b)

Fig. 1.1: (a) Recent refereed publications related to the field of self-healing materials, together with (b) their corresponding distribution of the employed keywords vocabulary. All published languages were included. All document types, including journal and conference articles, report papers, conference proceedings, and monograph published chapters were recorded. Statistics are available from 2000 to August 2010 inclusively. Data were collected from Engineering Village web-based information service, accessed from B. Aïssa, D. Therriault, E. Haddad, and W. Jamroz, "Self-Healing Materials Systems: Overview of Major Approaches and Recent Developed Technologies," Advances in Materials Science and Engineering, vol. 2012, Article ID 854203, 17 pages, 2012, accessed on 05/03/2019.

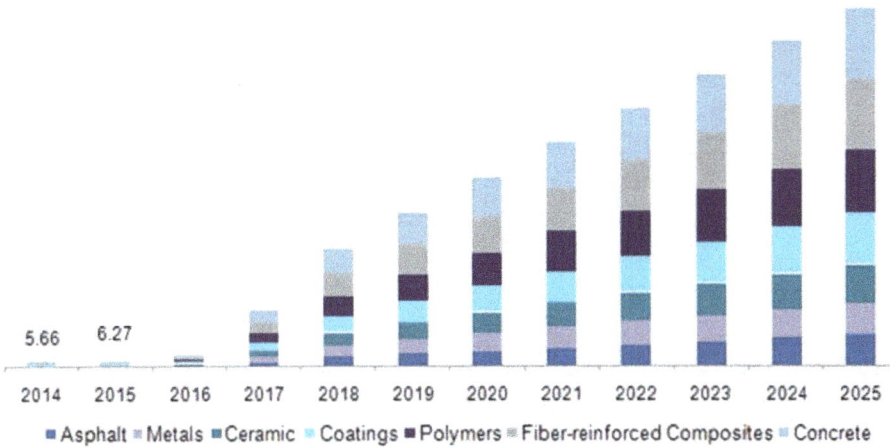

5.66   6.27

2014   2015   2016   2017   2018   2019   2020   2021   2022   2023   2024   2025

■ Asphalt ■ Metals ■ Ceramic ■ Coatings ■ Polymers ■ Fiber-reinforced Composites ■ Concrete

Fig. 1.2: U.S. self-healing materials market revenue (based on the product), 2014–2025 (in USD million) [23].

healing materials have grown due to the increasing global population, which induces demand for advanced infrastructural facilities. However, the high cost associated with such innovative and advanced products is predicted to emerge as a key restraint to the overall market development.

Various start-up companies in India and China are emerging in the field of self-healing polymers, especially in the paints and coating industry. The companies are focused on developing novel applications and cheaper products coupled with the commercialization of unique customized self-healing composites.

Following the application trends of the burgeoning self-healing market, building and construction showed the highest percentage share in 2016 with 27.4%. The report claims that the electronics and the telecommunication industry has witnessed a surprising growth of nearly 20.3% (Fig. 1.3). Emerging socioeconomic pressure to build hi-tech infrastructure and buildings in emerging economies of the Asia Pacific and Latin America is suggested to complement the development of the construction industry over the forecast period [26].

The third segment, which had seen a surge in the self-healing market, is the automobile industry. The original equipment manufacturers (OEMs) in the automotive sectors used self-healing coatings, which enhance the performance mechanisms of a cut or scratch. Decreasing the regular maintenance cost, this technology is projected to present an attractive market opportunity by pacing up the demands for advanced products in near future (Fig. 1.4).

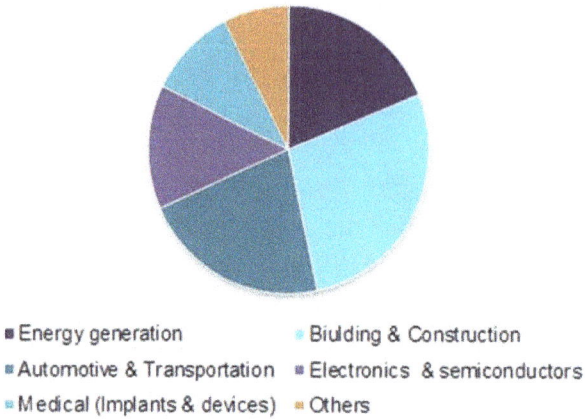

- Energy generation
- Biulding & Construction
- Automotive & Transportation
- Electronics & semiconductors
- Medical (Implants & devices)
- Others

Fig. 1.3: Global self-healing materials market revenue (%), by technology, 2016 [23].

GLOBAL SELF-HEALING MATERIAL MARKET

BY PRODUCT
- Concrete
- Coatings
- Polymers
- Metals
- Others

BY TECHNOLOGY
- Reversible Polymers
- Microenc apsulation
- Biological Material Systems
- Others

BY APPLICATION
- Construction
- Automotive
- Electronics
- Others

BY REGIONS
- North Americ a
- Europe
- Asia-Pacific
- Rest of the world

Fig. 1.4: Global trends of self-healing materials by product, application, and regions [85].

1.3 Introduction of self-healing materials

Self-healing is the property of exhibiting the phenomenon to rejoin the fractured/damaged part of a material autonomously. In due course of time, materials degrade due to environmental parameters, causing damages and defects like fatigue, cracks, and tear [27] on a microscopic level. These microstates of the fracture can spread throughout the material eventually yielding a brittle and ductile failure. Cracks, especially in their microscopic state are very difficult to detect, and thus they need a rhetorical system checkup. To bypass or eliminate the entire checking process, smart polymers, metals, and composites have been developed so that the cracks heal by themselves when subjected to an external stimulus [28]. Although we are most familiar with polymers behaving like self-healing materials because of their highly entangled structure, there are various metals, ceramics, and cementitious materials displaying similar attributes.

1.3.1 Self-healing fiber/polymer composites

External systems for the fabrication of self-healing materials include discrete capsule-based development and continuous vascular systems [29]. Researches are still on to illuminate the field of intrinsically healed fiber/polymer composites [30]. The application of the extrinsic system remains limited since there are a number of healing instances where we need a matrix based on intrinsically healing agents. T-joints and aircraft fuselages use the above technologies of both capsule and vascular systems to inherit the property of self-healing [31].

1.3.2 Self-healable polymer coatings

Various researches haves shown that self-healing materials can be used as coating layers in various ways such as:
a) Microencapsulation [32]
b) Reversible physical bonds [33]
c) Ionomers [34]
d) Chemical bonds [35]

The microencapsulation technique was one of the prime approaches to synthesize self-healing coatings with the introduction of the dicyclopentadiene (DCPD) and Grubb's catalyst in an epoxy matrix [36]. In sensible electronic applications, liquid metal microdroplets are being used with silicone elastomers in order to impart flexibility to the material. New innovation suggests that apart from DCPD, several other materials do display self-healing behavior when incorporated into suitable matrices;

for example, isocyanate, (glycerol methacrylate) GMA, linseed oil, and Tung oil [37, 38], which can be applied as a self-healing coating material (Fig. 1.5).

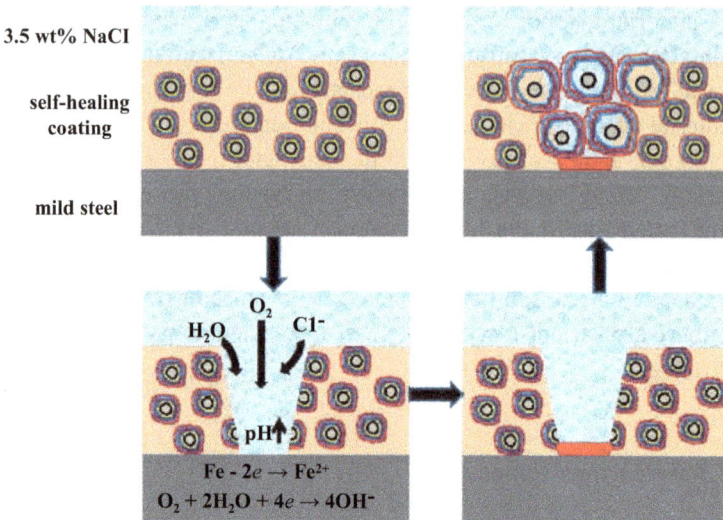

3.5 wt% NaCl

self-healing coating

mild steel

O_2
H_2O Cl^-

$pH\uparrow$

$Fe - 2e \rightarrow Fe^{2+}$
$O_2 + 2H_2O + 4e \rightarrow 4OH^-$

Fig. 1.5: Elucidating the self-healing property as a coating material [86].

1.3.3 Self-healing cementitious materials

The first reported self-healing behavior from a cementitious material was in 1836 by the French Academy of Sciences [39]. Since then, various technologies have been discovered that aim to upgrade the physical and mechanical properties of these materials by a chemical or a biochemical strategy.

a. Autonomous healing is the natural ability of the cementitious material to heal with the application or the induction of any external stimulus. The most accepted way through which the reaction proceeds is by the hydration of the un-hydrated cement particles followed by the carbonation of the dissolved calcium hydroxide [40].

b. Chemical agents such as the strategies implemented in the use of the capsular and vascular process can be used to self-heal cementitious materials [41]. The capsules or the vesicles once ruptured can heal the cracks that are up to 0.2 mm thick [42]. Fig. 1.6 shows self-healing process in a three-component system (water/ binder mineral admixture). Here Ahn and Kishi showcased the phase wise healing of a crack with an initial width of 0.2 mm. Rehydration products between the cracks were gradually observed after 14 days which got healed almost perfectly after 200 days with some minute cracks in between [87]

c. Biologically enhanced self-healing stimulus can also be implemented to fetch the desired self-repairable properties. The incorporation of bacteria, which can pro-

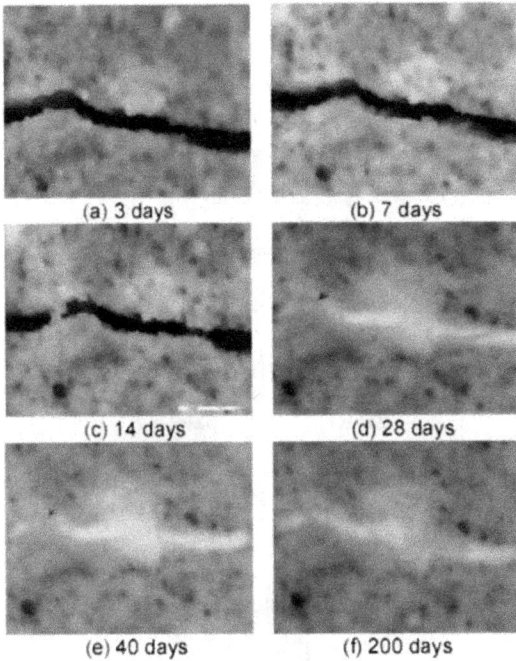

(a) 3 days	(b) 7 days
(c) 14 days	(d) 28 days
(e) 40 days	(f) 200 days

Fig. 1.6: Investigating the property of self-healing in cementitious materials [88].

mote calcium carbonate precipitation via their metabolic pathways, can be used to initiate the process of self-healing [43]. This precipitate can not only form the newly synthesized healing material but also acts as a layer with superior barrier properties.

1.3.4 Self-healing ceramic–polymer hybrids

Being superior in strength in the high-temperature application, ceramics are advantageous in the sense that they can be used in high-temperature applications. The only disadvantage is that the material is brittle and susceptible to crack formation [44]. Max phase's ceramics have the ability to heal the cracks via intrinsic healing mechanisms. The oxides liberated by the max phase ceramics fill the void thus rephrasing the cracks efficiently [45]. Another approach to self-heal a ceramic material is by embedding certain metals like alumina or zirconium in a matrix, whereby when exposed to heat or oxygen they react to form their respective oxides, which indeed complements the self-healing process [46]. At higher temperatures, SiC has proven to be a potential ceramic material in applications to restore the fractured surface (Fig. 1.7) [47].

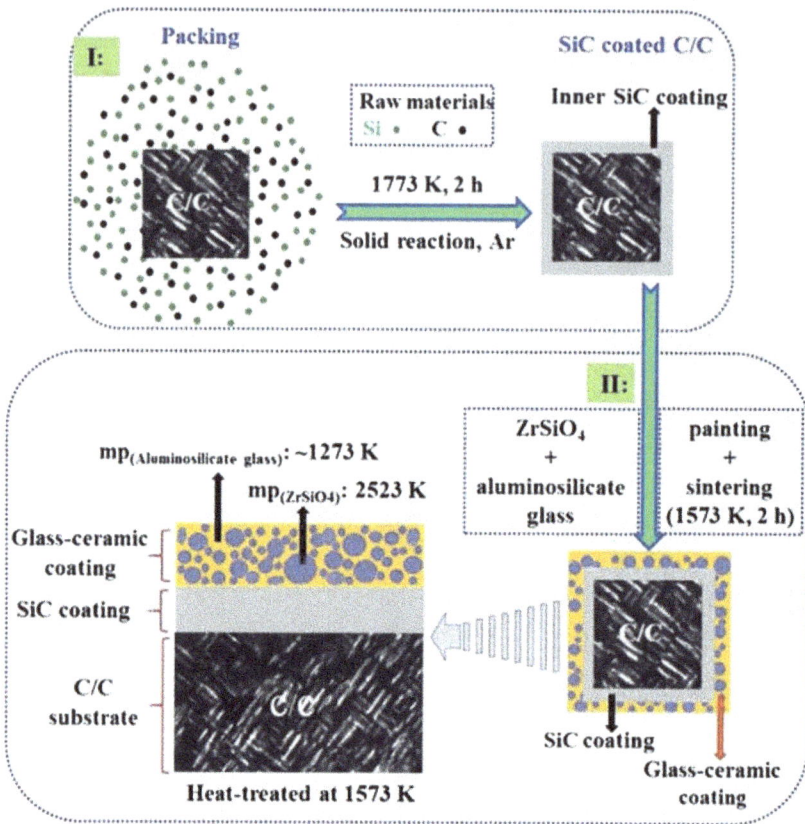

Fig. 1.7: Illustration of self-healing property in ceramic materials [89].

1.3.5 Self-healing metal–polymer hybrids

Metals that exhibit premature and low ductility creep fracture due to exposure to temperatures can be healed with the application of smart materials [48]. These small fracture defects amalgamate to form a crack resulting in a macroscopic failure. Due to the high melting point and low molecular mobility, achieving the self-healing phenomenon through the intrinsic method is difficult [10]. Precipitation is the primary way by which we can make metals exhibit the property of self-regeneration in metals. Research conducted with copper and gold shows that metals can efficiently be healed when subjected to a stimulus [49]. The healing agents selectively precipitate at the voids of the creep cavity, yielding in pore filling and closure. Reports extend that for lower stress levels up to 80% filling of the creep cavities can be healed when subjected to precipitated gold (Fig. 1.8) [50–52].

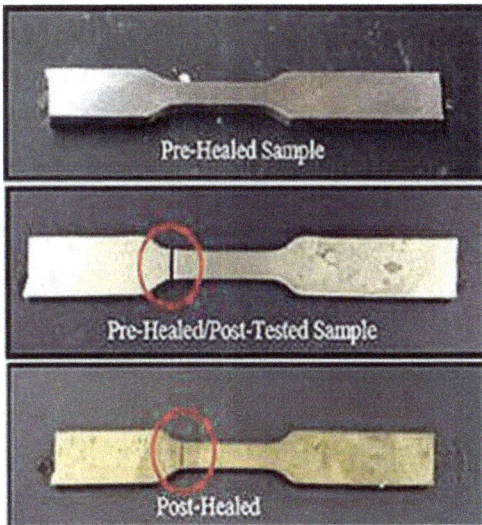

Fig. 1.8: Demonstration of self-healing effect in metals [90].

1.4 Introduction of elastomer and thermoplastic elastomer (TPE)-based self-healing compounds

Recently, elastomers have been investigated in the various realms of engineering such as tires, shoes, and seals [53–56]. An elastomer is classified as a viscoelastic material with weak intermolecular forces and a relatively lower modulus along with a higher failure strain compared to plastics [57, 58]. The resistance to deformation originates primarily from the entropy. Thus, when entropy drops, the free energy increases and external stress is applied to impose the deformation [59]. The first self-healing elastomer was incepted in 2008 [60] and since then the research on self-healing elastomers has escalated; primarily they can meet various sustainability demands. Nowadays, there are various chemistries that are employed to fabricate robust self-healing elastomers based on various reversible and irreversible covalent bonds, supramolecular chemistry, chemo-mechanics, and shape-memory-assisted self-healing [59]. Considering the issue of sustainability, elastomers often turn out to be a disadvantage owing to their permeant cross-linking that prevents reusing them easily [61, 62]. Moreover, the increasing demand for rubber products causes rubber waste challenging the research community to come up with a solution.

With recent advancements, there are two approaches to address the aforementioned problem. First, to come up with a strategy to develop a re-processable cross-linked elastomer and second to improve the shelf-life of the elastomer by modifying them to self-healable [63, 64]. The self-healable elastomer has gained interest owing to no need for replacement, and we see this effort being promoted every year with the increase in the number of literature along with the growth in the self-healing elasto-

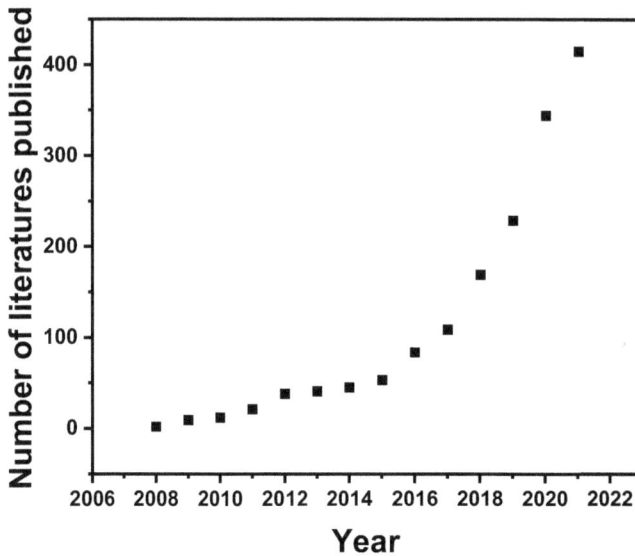

Fig. 1.9: The number of literature published with the search term "self-healing elastomer," data sourced from Scifinder, 2022.

mer industry (Fig. 1.9). When a crack is generated in the elastomer matrix, the elastomer autonomously repairs the crack when stimulated with an external stimulus, thereby improving its application life and a reduction in the wastes generated.

Since a majority of the elastomer application traverses to the tire industry, conjugated diolefin-based elastomers (polybutadiene, styrene-butadiene rubber, polyisoprene, butyl rubber, nitrile butadiene rubber, and chloroprene rubber) are extensively used to generate self-healing elastomers. Recent trends have also seen a rise in the development of functional elastomer and functional elastomers based on dynamic covalent networks or supramolecular interactions, which help in the further reprocessing of the elastomer [65, 66].

Thermoplastic elastomers (TPEs) are the class of polymers that combine the properties of both rubber and thermoplastics in terms of elastic behavior, service properties, and processability, respectively. Their continuous development and growth have made TPEs the most rapidly growing segment of polymer technology with significant commercial importance. Processing feasibility, recyclability, tailoring in structure and properties, and other several advantages over conventional thermoset rubbers coupled with a wide selection of performance capabilities are the key factors responsible for their sustained growth and continued acceptance in the commercial field [67]. In short, TPE contains biphasic morphological structure having two different kinds of structural units, namely, a hard crystalline segment with a high glass transition temperature (T_g) and a soft amorphous component having low T_g. The hard segment acts as thermolabile physical cross-links whereas the soft segments are responsible for the dynamic movement of polymer chains that allow TPEs to soften and flow at elevated temperatures. But thermodynamic incompatibility must be present between the two segments at service temperature to restrain interpenetration of the segments [68].

In the context of self-healing application, TPEs have been proved to be an ideal material as they can be easily melted, processed, and recycled offering a wide range of physical and chemical properties. Initially achieving mechanical strength and stiffness along with rapid dynamic molecular motion simultaneously in a single material was the key challenge as well as a major hindrance in fabricating self-healing material [69]. Multiphase design of TPEs addressed the dilemma of two contradictory properties of robustness and self-healing efficiency. Generally, TPEs are block or brush copolymers comprising hard glassy domains embedded in soft rubbery segments and the hard domains provide high modulus and stiffness apart from being the physical cross-links, while soft segment leads to the interdiffusion of polymer chains above transition temperature [70]. Localized segmental mobility around the wounded area and synchronized recombination of macromolecular reactive segments into the voids or free volume areas are the two critical phenomena during the self-mending process, which may occur continuously depending upon the kinetics and thermodynamics interplay over there. Heterogeneous network like TPE comprises spatially distributed hard and soft segments, which can rearrange themselves upon cleavage to facilitate healing [71].

Covalent connectivity of the soft segments between hard domains precludes healing below T_g or melting temperature (T_m) of the hard phase. Initiated by this seminal report, scientists have adopted many exciting strategies to replace those covalent link-

Tab. 1.1: Self-healing attributes in some commercially used TPEs.

Polymer	Reinforcement	Self-healing moiety	Nature of interaction	Stimuli	Reference
Polystyrene-b-polybutadiene-b-polystyrene (SBS) and polystyrene-b polyisoprene-b-polystyrene (SIS)	–	Mutual diffusion of block components	No chemical reaction	Room temperature	[73]
Brush copolymers having glassy polymethylmethacrylate (PMMA) backbone and flexible polyacrylate-amide (PA-amide) brushes	–	Secondary amide groups in soft PA-amide phase	Dynamic hydrogen bond	Room temperature	[74]

Tab. 1.1 (continued)

Polymer	Reinforcement	Self-healing moiety	Nature of interaction	Stimuli	Reference
TPE based on oligomeric charged triblock copolymer containing poly(2-(dimethylamino) ethyl acrylate) (PDMAEA)-*b*-poly (nbutylacrylate) (PBA)-*b*-poly (2-carboxyethylacrylate (PCEA)	–	Block components	Supramolecular network based on electrostatic interaction	Mild heating	[75]
Hydrogen bonded brush copolymer containing hard polystyrene backbone with soft polyacrylate amide (PA-amide)	–	PA-amide	Supramolecular assembly due to H-bonding	Ambient temperature, autonomous	[70]
Thermoplastic polyurethane (TPU)	Multiwall carbon nanotube (MWCNT)	TPU itself	Bridging tendency of CNT, assisting diffusion and re-entanglement of TPU	UV light	[76]

ages by reversible covalent interactions like cycloaddition reaction, metathesis or re-shuffling reaction, and non-covalent interactions namely H-bonding, metal–ligand interactions, π–π stacking, and ionomers based on supramolecular structure [70, 72] which will be elucidated in detail in the subsequent chapters. Self-healing attributes of some TPEs have been summarized in Tab. 1.1 for further reference.

The intrinsic self-healing ability of a material is driven by intra- and intermolecular interactions due to the presence of dynamic reversible bonds that undergo exchange reactions. The process of self-healing consists of two vital steps, which are: a) closure of the cut/scratch and b) sealing of the gap. The first step is dependent on the viscoelastic property of the material, while the later is on the local dynamic process by means of polymer interdiffusion and bond restoration to reestablish the network. Self-healing, being exclusively a diffusion-driven phenomenon, the rheological aspect has become one of the key factors to impact the entire process of healing, and various rheological measurements are performed to monitor these exchange reactions that significantly affect the microstructural changes in polymers [77–80]. Supramolecular groups in functional smart materials like self-healing are responsible for the reversibility of secondary interactions, and they are studied intensively by rheological methods to explore the material's response to mechanical stimuli; in particular, their mechanical relaxation behavior plays an important role. Relaxation rate is directly related to molecular mobility that contributes towards macroscale mobility and consequently self-healing. The rheological relaxation pattern is significantly influenced by the reassociating groups, which can be introduced in the materials by backbone functionalization capable of forming different dynamic reversible bonds. Strong intermolecular association in the system increases the terminal relaxation time imposing configurational restrictions on long-chain rearrangements, which reflects in the viscoelastic response of polymers [81].

For example, when comparing copolymers synthesized from n-butyl acrylate (nBA) and various functional comonomers such as ureidopyrimidinone (UPy)-functionalized acrylate, acrylamidopyridine, acrylic acid, or carboxyethyl acrylate, a noticeable difference is observed in their hydrogen-bonding dynamics. In systems where the incorporated functional groups form relatively weak hydrogen bonds, the rate of bond dissociation and reformation is faster than the characteristic timescale of chain relaxation. This behavior resembles that of unentangled polymer melts, where the macromolecular chains exhibit high segmental mobility and limited elastic behavior.

In contrast, copolymers bearing strong hydrogen-bonding moieties such as UPy groups, which form stable dimeric complexes, exhibit markedly different rheological properties. The presence of such strong and long-lived hydrogen bonds (with dimer lifetimes exceeding 10 s) promotes the formation of a transient but physically entangled network. This network behaves as an elastic solid over relevant timescales due to the restriction of chain mobility. Furthermore, these strong interactions increase the flow activation energy, indicating that more thermal energy is required to induce flow in the system [82].

A similar trend is observed in polymers based on poly(2-ethylhexyl methacrylate) (PEHMA) copolymerized with UPy-functionalized methacrylates. As the content of the UPy units increases, both the complex viscosity and the duration of the plateau in the storage modulus (G') region increase, consistent with the formation of a more robust transient network. In another related example, copolymers incorporating 7.2 wt% of UPy-acrylate demonstrated full recovery (up to 100%) of self-adhesion strength on fractured surfaces after 50 h of healing, indicating effective supramolecular self-healing behavior under ambient conditions [83].

However, it is important to note that self-healing efficiency is not governed by supramolecular interactions alone. The rheological properties and the inherent mobility of the polymer chains play crucial roles in dictating the rate and extent of the healing process. For instance, materials with high chain mobility and lower glass transition temperatures (T_g) tend to allow better segmental rearrangement at damaged interfaces, facilitating effective healing. Conversely, materials with highly viscous or glassy matrices may exhibit sluggish or incomplete healing due to the limited movement of macromolecular chains.

In addition to supramolecular systems, dynamic covalent networks and associative reversible bonding chemistries also influence rheological behavior. In comparison to permanent covalent C–C bonds, non-covalent dynamic interactions such as hydrogen bonds, π–π stacking, or metal–ligand coordination enable time-dependent relaxation through bond exchange or dissociation–association events. These mechanisms enhance the chain mobility and allow stress relaxation even in cross-linked materials. At elevated temperatures, such non-covalent systems can undergo structural rearrangement, leading to macroscopic flow, although the timescales involved are considerably longer than in linear systems. This behavior contrasts with traditional covalently cross-linked thermosets, where the permanent network inhibits any form of viscoelastic flow, rendering the material rigid and brittle under stress [84].

To model these observations quantitatively, various rheological frameworks have been developed that account for the reversible nature of physical bonds, including transient network theory, sticky Rouse models, and generalized Maxwell models adapted for dynamic bonding. These models attempt to relate microscopic bonding lifetimes and association–dissociation kinetics to the macroscopic viscoelastic response of the material.

In this book, we aim to systematically analyze the rheological properties of such dynamic polymer networks, beginning from the molecular origin of supramolecular interactions and extending to their macroscopic implications. Particular emphasis is placed on understanding how the strength, density, and dynamics of these reversible bonds influence rheological parameters such as complex viscosity, storage/loss moduli, and relaxation time. Furthermore, these rheological insights are directly correlated with practical phenomena such as scratch resistance and self-healing behavior. In the subsequent chapters, detailed case studies are presented to illustrate these relationships, supported by both experimental data and theoretical analysis.

References

[1]   White, R., S,, Sottos, N., Geubelle, P., Moore, J., Kessler, R., Sriram, M., Suresh,, Brown, E. & Viswanathan, S. (2001). Autonomic healing of polymer composites. *Nature*, *409*, 794–797.
[2]   Zhang, B., Zhang, P., Zhang, H., Yan, C., Zheng, Z., Wu, B. & Yu, Y. (2017). A transparent, highly stretchable, autonomous self-healing Poly(dimethyl siloxane) elastomer. *Macromolecular Rapid Communications*, *38*.
[3]   Canadell, J., Goossens, H. & Klumperman, B. (2011). Self-healing materials based on disulfide links. *Macromolecules*, *44*, 2536–2541.
[4]   Xu, Y. & Chen, D. (2016). A Novel Self-Healing Polyurethane Based on Disulfide Bonds. *Macromolecular Chemistry and Physics*, *217*, 1191–1196.
[5]   Tuncaloylu, D. C., Sau, M. & Opperman, W. (2011). Tough and Self-healing hydrogels formed via hydrophobic interaction. *Macromolecules*, *44*, 4997–5005.
[6]   Luo, F., Sun, T. L., Nakajima, T., Kurokawa, T., Ihsan, A. B., Xufeng, L., Guo, H. & Gong, J. P. (2015). Reprocessability of Tough and Self-Healing Hydrogels Based on Polyion Complex. *ACS Macro Letters*, *4*(9), 961–964.

[7]   Novoselow, K. S., Guin, A. K. & Jiang, D. (2004). Electric field effect in automatically thin carbon films. *Science*, *306*, 666–669.

[8]   Mozhdehi, D., Ayala, S., Cromwell, O. R. & Guan, Z. (2014). Self-Healing Multiphase Polymers via Dynamic Metal–Ligand Interactions. *Journal of the American Chemical Society*, 136(46), 16128–16131.

[9]   Bucknall, C. B., Drinkwater, I. C. & Smith, G. R. (1980). Hot plate welding of plastics: factors affecting weld strength. *Polymer Engineering & Science*, 20(6), 432–440.

[10] Pang, J. W. C. & Bond, I. P. (2005). Bleeding composites – damage detection and self-repair using a biomimetic approach. *Composites: Part A Applied Science and Manufacturing*, 36(2), 183–188.

[11] Varghese, S., Lele, A. & Mashelkar, R. (2006). Metal-ion-mediated healing of gels. *Journal of Polymer Science Part A: Polymer Chemistry*, *44*, 666–670.

[12] Wayman, E. (2011). The Secrets of Ancient Rome's Buildings. *Smithsonian*, Accessed on 5th March, 2019.

[13] Back to the Future with Roman Architectural Concrete, *Lawrence Berkeley National Laboratory*, The University of California, 2014. Accessed on 5th March, 2019.

[14] Hartnett, K. (2014). Why is ancient Roman concrete still standing, *Boston Globe*, Accessed on 5th March, 2019.

[15] Jackson Marie, D., Landis Eric, N., Brune Philip, F., Massimo, V., Heng, C., Qinfei, L., Martin, K., Hans-Rudolf, W., Monteiro Paulo, J. M. & Ingraffea Anthony, R. (2013). Mechanical resilience and cementitious processes in Imperial Roman architectural mortar. *Proceedings of the National Academy of Sciences*, *111(52)*, 18484–18489.

[16] Liu, D., Lee, C. Y. & Lu, X. (1993). Repairability of impact-induced damage in SMC composites. *Journal of Composite Materials*, *27*(13), 1257–1271.

[17] Osswald, T. & Menges, G. (2012). Failure and damage of polymers. Materials Science of Polymers for Engineers. In Hanser, Munich, Germany, pp. -447.

[18] Jud, K., Kausch, H. H. & Williams, J. G. (1981). Fracture mechanics studies of crack healing and welding of polymers. *Journal of Materials Science*, *16*(1), 204–210.

[19] Dry, C. M. & Sottos, N. R., Passive smart self-repair in polymer matrix composite materials. In Proceedings of the Conference on Recent Advances in Adaptive and Sensory Materials and Their Applications, Technomic: Virginia, USA, 1993, 438–444.

[20] White, S. R., Sottos, N. R., Geubelle, P. H. et al. (2001). Autonomic healing of polymer composites. *Nature*, *409*(6822), 794–797.

[21] Carlson, H. C. & Goretta, K. C. (2006). Basic materials research programs at the U.S. Air Force Office of Scientific Research. *Materials Science and Engineering B: Solid-State Materials for Advanced Technology*, *132*, 1–2, 2–7.

[22] Semprimosching, C. (2006). Enabling self-healing capabilities – a small step to bio-mimetic materials. In *Tech. Rep*, European Space Agency Materials: Noordwijk, The Netherlands, Vol. 4476.

[23] https://www.grandviewresearch.com/industry-analysis/self-healing-materials, Accessed on 5th March, 2019

[24] Gluson, B. M. (1960). *Polymer*, *1*, 499.

[25] Edgar, O. B. & Hill, R. (1952). The *p*-phenylene linkage in linear high polymers: Some structure-property relationships. *Journal of Polymer Science*, *8*, 1–22.

[26] Meng, H. & Hu, J. L. (2010). A brief review of stimulus-active polymers responsive to thermal, light, magnetic, electric and water/solvent stimuli. *Journal of Intelligent Material Systems & Structures*, *21*, 859–885.

[27] Leng, J., Lu, H., Liu, Y., Huang, W. M. & Du, S. (2009). *MRS Bulletin*, *34*, 848–855.

[28] Wagermaier, W., Kratz, K., Heuchel, M. & Lendlein, A. (2010). Characterization methods for shape-memory polymers. *Advances in Polymer Science*, *21*, 654–661.

[29] Yang, I., Zhang, J. & Wang, C. (2013). Healing of Carbon fiber epoxy composites T joints using mendable polymer fiber stitching. *Composites B*, *45*, 1499–1504.

[30]   Lendlein, A. & Kelch, S. (2005). Shape-memory polymers as stimuli-sensitive implant materials. *Clinical Hemorheology and Microcirculation, 32*(2), 105–116.

[31]   Guignot, C., Betz, N., Legendre, B., Le Moel, A. & Yagoubi, N. (2001). Degradation of segmented poly(etherurethane) Tecoflex® induced by electron beam irradiation: characterization and evaluation. *Nuclear Instruments and Methods in Physics Research Section B, 185*, 100–107.

[32]   Mohr, R., Kratz, K., Weigel, T., Lucka-Gabor, M., Moneke, M. & Lendlein, A. (2006). Initiation of shape-memory effect by inductive heating of magnetic nanoparticles in thermoplastic polymers. *Proceedings of the National Academy of Sciences of the United States of America, 103*(10), 3540–3545.

[33]   Yakacki, C. M., Shandas, R., Safranski, D., Ortega, A. M., Sassaman, K. & Gall, K. (2008). Strong, tailored, biocompatible shape-memory polymer networks. *Advanced Functional Materials, 18*(16), 2428–2435.

[34]   Chem, X. & guan, Y. (2012). Multivalent hydrogen bonding block copolymers self assemble into strong, tough self-healing materials. *Chemicals Communication, 50*, 10868–10870.

[35]   Lendlein, A. & Langer, R. (2002). Biodegradable, elastic shape-memory polymers for potential biomedical applications. *Science, 296*(5573), 1673–1676.

[36]   Meng, Q. & Hu, J. (2008). Self-organizing alignment of carbon nanotube in shape memory segmented fiber prepared by in situ polymerization and melt spinning. *Composites Part A: Applied Science and Manufacturing, 39*, 314–321.

[37]   Kim, J. Y., Han, S. I. & Hong, S. (2008). Effect of modified carbon nanotube on the properties of aromatic polyester nanocomposites. *Polymer, 49*, 3335–3345.

[38]   Eric, J. & Michael, D. (2017). An autonomously electrically self-healing liquid metal elastomeric composite for robust soft robotics. *Nature Materials*, 1476-1122(165N).

[39]   Rooij, M. & Van, K. (2013). Self-healing phenomenon in cement-based materials. In Otten, E. (ed), Springer Netherlands. 6624-2, pp. ISBN-978-94007.

[40]   Edvardsen, C. (1999). Water permeability and autogenous healing of cracks in concrete. *ACI Materials Journal, 96*, 448–454.

[41]   Snoeck, D., De Schryver, T. & De Belie, N. (2018). Enhanced impact energy absorption in self-healing strain-hardening cementitious materials with superabsorbent polymers. *Construction and Building Materials, 191*, 13–22.

[42]   Snoeck, D. (2018). Superabsorbent polymers to seal and heal cracks in cementitious materials. *RILEM Technical Letters, 3*, 32.

[43]   Tsardaka, E.-C., Stefanidou, M. & Pavlidou, E. (2018). The role of nanoparticles on the durability of lime-pozzolan binding system, Accessed on 5th March, 2019.

[44]   Ono, M., Nakao, W., Takahashi, K., Nakatani, M. & Ando, K. (2007). A New Methodology to Guarantee the Structural Integrity of Al_2O_3/SiC Composite Using Crack Healing and a Proof Test. *Fatigue and Fracture of Engineering Materials and Structures, 30*, 599–607.

[45]   Ozaki, S., Aoki, Y., Osada, T., Takeo, K. & Nakao, W. (2018). Finite element analysis of fracture statistics of ceramics: Effects of grain size and pore size distributions. *Journal of the American Ceramic Society*, Accessed on 5th March, 2019.

[46]   Boatemaa, L., Kwakernaak, C., Zwaag, S. & Sloof, W. G. (2016). Selection of healing agents for autonomous healing of alumina at high temperatures. *Journal of the European Ceramic Society, 36*.

[47]   Takeda, K., Tanahashi, M. & Unno, H. (2003). Self-repairing mechanism of plastics. *Science and Technology of Advanced Materials, 4*, 435–444.

[48]   Pang, J. W. C. & Bond, I. P. (2005). A hollow fiber reinforced polymer composite encompassing self-healing and enhanced damage visibility. *Composites Science & Technology, 65*, 1791–1799.

[49]   Y-l., L. & Chen, Y.-W. (2007). Thermally reversible crosslinked polyamides with high toughness and self-repairing ability from maleimide- and furan-functionalized aromatic polyamides. *Macromolecular Chemistry and Physics, 208*, 224–232.

[50] Dry, C. (1994). Matrix cracking repair and filling using active and passive modes for smart timed release of chemicals from fibers into cement matrices. *Smart Materials and Structures, 3*, 118–123.

[51] Wool, R. P. (2008). Self-healing materials. *Soft Matter, 4*, 400–418.

[52] Wu, S., Li, J., Zhang, G., Yao, Y., Li, G., Sun, R., Wong, C. (2017). Ultrafast Self-Healing Nanocomposites via Infrared Laser and Their Application in Flexible Electronics. *ACS Applied Materials & Interfaces, 9* (3).

[53] Attaran, S. A., Hassan, A. & Wahit, M. U. (2017). Materials for food packaging applications based on bio-based polymer nanocomposites: A review. *Journal of Thermoplastic Composite Materials, 30*(2), 143–173.

[54] Adhikari, B., De, D. & Maiti, S. (2000). *Progress in Polymer Science, 25*, 909–948.

[55] Bockstal, L., Berchem, T., Schmetz, Q. & Richel, A. (2019). Devulcanisation and reclaiming of tires and rubber by physical and chemical processes: A review. *Journal of Cleaner Production, 236*, 117574.

[56] Morin, J. E., Williams, D. E. & Farris, R. J. (2002). A novel method to recycle scrap tires: high-pressure high-temperature sintering. *Rubber Chemistry and Technology*, 75(5), 955–968.

[57] Zuo, X.-L. et al. (2021). Self-healing polymeric hydrogels: toward multifunctional soft smart materials. *Chinese Journal of Polymer Science, 39*(10), 1262–1280.

[58] Wang, D., Wang, Z., Ren, S., Xu, J., Wang, C., Hu, P. & Fu, J. (2021). Molecular engineering of a colorless, extremely tough, superiorly self-recoverable, and healable poly (urethane–urea) elastomer for impact-resistant applications. *Materials Horizons, 8*(8), 2238–2250.

[59] Wang, Z., Lu, X., Sun, S., Yu, C. & Xia, H. (2019). Preparation, characterization and properties of intrinsic self-healing elastomers. *Journal of Materials Chemical B, 7*(32), 4876–4926.

[60] Cordier, P., Tournilhac, F., Soulie-Ziakovic, C., Leibler, L. (2008). *Nature, 451*, 977–980.

[61] Imbernon, L. & Norvez, S. (2016). From landfilling to vitrimer chemistry in rubber life cycle. *European Polymer Journal, 82*, 347–376.

[62] Hernández Santana, M., Den brabander, M., García, S. & Van der zwaag, S. (2018). Routes to make natural rubber heal: a review. *Polymer Reviews, 58*(4), 585–609.

[63] White, S. R., Sottos, N. R., Geubelle, P. H., Moore, J. S., Kessler, M. R., Sriram, S. R. . . . Viswanathan, S. (2001). Autonomic healing of polymer composites. *Nature, 409*(6822), 794–797.

[64] Samadzadeh, M., Boura, S. H., Peikari, M., Kasiriha, S. M. & Ashrafi, A. (2010). A review on self-healing coatings based on micro/nanocapsules. *Progress in Organic Coatings, 68*(3), 159–164.

[65] Wu, X. F., Rahman, A., Zhou, Z., Pelot, D. D., Sinha-Ray, S., Chen, B. . . . Yarin, A. L. (2013). Electrospinning core-shell nanofibers for interfacial toughening and self-healing of carbon-fiber/epoxy composites. *Journal of Applied Polymer Science, 129*(3). 1383–1393.

[66] Utrera-Barrios, S., Verdejo, R., López-Manchado, M. A. & Santana, M. H. (2020). Evolution of self-healing elastomers, from extrinsic to combined intrinsic mechanisms: A review. *Materials Horizons, 7*(11), 2882–2902.

[67] Kear, K. E. (2003). Introduction. In: Humphreys, D. S. (ed.). Developments in Thermoplastic Elastomers, RAPRA, U.K, pp. 3–11.

[68] Costa, F. R., Dutta, N. K., Choudhury, N. R. & Bhowmick, A. K. (2008). Thermoplastic Elastomers. In: Bhowmick, A. K. (ed.), Current Topics in Elastomers Research, Taylor & Francis, Boca Raton, pp. 101–164.

[69] Hentschel, J., Kushner, A. M., Ziller, J. & Guan, Z. (2012). Self-Healing Supramolecular Block Copolymers. *Angewandte Chemie, 124*, 10713–10717.

[70] Chen, Y., Kushner, A. M., Williams, G. A. & Guan, Z. (2012). Multiphase design of autonomic self-healing thermoplastic elastomers. *Nature Chemistry, 4*, 467–472.

[71] Yang, Y., Ding, X. & Urban, M. W. (2015). Chemical and physical aspects of self-healing materials. *Progress in Polymer Science, 49–50*, 34–59.

[72] Yang, J.-X., Long, Y.-Y., Pan, L., Men, Y.-F. & Yue-Sheng, L. (2016). Spontaneously Healable Thermoplastic Elastomers Achieved through One-Pot Living Ring-Opening Metathesis Copolymerization of Well-Designed Bulky Monomers. *Applied Materials & Interfaces, 8*, 12445–12455.

[73] Watanabe1, R., Sako1, T., Korkiatithaweechai1, S. & Yamaguchi, M. (2017). Autonomic healing of thermoplastic elastomer composed of triblock copolymer. *Journal of Materials Science, 52,* 1214–1220.

[74] Chen, Y. & Guan, Z. (2015). Self-healing thermoplastic elastomer brush copolymers having a glassy polymethylmethacrylate backbone and rubbery polyacrylate-amide brushes. *Polymer, 69,* 249–254.

[75] Voorhaar1, L., Diaz, M. M., Leroux, F., Rogers, S., Abakumov, A. M., Van Tendeloo, G., Van Assche, G., Van Mele, B. & Hoogenboom, R. (2017). Supramolecular thermoplastics and thermoplastic elastomer materials with self-healing ability based on oligomeric charged triblock copolymers. *Supramolecular polymeric materials, 9,* 1–10.

[76] Deshmukh, K., Bhakare, M., Shrivastav, S., Naik, V., Swamini Chopra, C. L. G. & Peshwe, D. R. (2017). Synthesis and Characterization of Self Healing Thermoplastic Polyurethane (TPU) Thin Films Reinforced with Multi-wall Carbon Nanotubes (MWCNTs). *Journal of Nano Technology and Its Application in Engineering, 2,* 1–13.

[77] Wang, Z., Zhou, J., Liang, H., Ye, S., Zou, J. & Yang, H. (2020). A novel polyurethane elastomer with super mechanical strength and excellent self-healing performance of wide scratches. *Progress in Organic Coatings, 149,* 105943.

[78] Eom, Y., Kim, S. M., Lee, M., Jeon, H., Park, J., Lee, E. S. . . . Oh, D. X. (2021). Mechano-responsive hydrogen-bonding array of thermoplastic polyurethane elastomer captures both strength and self-healing. *Nature communications, 12*(1), 1–11.

[79] Lee, S. H. & Lee, D. S. (2019). Self-healing and rheological properties of polyhydroxyurethane elastomers based on glycerol carbonate capped prepolymers. *Macromolecular Research, 27*(5), 460–469.

[80] Hu, J., Mo, R., Sheng, X. & Zhang, X. (2020). A self-healing polyurethane elastomer with excellent mechanical properties based on phase-locked dynamic imine bonds. *Polymer Chemistry, 11*(14), 2585–2594.

[81] Gold, B. J., Hövelmann, C. H., Lühmann, N., Pyckhout-Hintzen, W., Wischnewski, A. & Richter, D. (2017). The microscopic origin of the rheology in supramolecular entangled polymer networks. *Journal of Rheology, 61*(6), 1211–1226.

[82] Lewis, C. L., Stewart, K. & Anthamatten, M. (2014). The influence of hydrogen bonding side-groups on viscoelastic behavior of linear and network polymers. *Macromolecules, 47*(2), 729–740.

[83] Enke, M., Döhler, D., Bode, S., Binder, W. H., Hager, M. D. & Schubert, U. S. (2015). *Intrinsic Self-Healing Polymers Based on Supramolecular Interactions: State of the Art and Future Directions. Advances in Polymer Science, 59–112.* doi: 10.1007/12_2015_345.

[84] Nevejans, S., Ballard, N., Fernández, M., Reck, B., García, S. J. & Asua, J. M. (2019). The challenges of obtaining mechanical strength in self-healing polymers containing dynamic covalent bonds. *Polymer, 179,* 121670.

[85] https://www.marketresearchfuture.com/reports/self-healing-materials-market-5503

[86] Bei, Q. & S., Zuwei &Hao, L., Wang, W. & Kong, D. (2017). Self-Healing Epoxy Coatings Based on Nanocontainers for Corrosion Protection of Mild Steel. *Journal of The Electrochemical Society, 164,* C54–C60. doi: 10.1149/2.1251702jes.

[87] T.H. Ahn, T. Kishi, Crack self-healing behavior of cementitious composites incorporating various mineral admixtures, J. Adv. Concr. Technol. 8(2) (2010) 16.

[88] Huang, H., Ye, G., Qian, C. & Schlangen, E. (2016). Self-healing in cementitious materials: Materials, methods and service conditions. *Materials and Design, 92,* 499–511.

[89] Zhen, Q., Zheng, L., Pengfei, H., Song, S. & Zheng, F. (2018). A glass-ceramic coating with self-healing capability and high infrared emissivity for carbon/carbon composites. *Corrosion Science, 141,* 81–87.

[90] http://brinson.mech.northwestern.edu/research/SMA_Composites.html

Chapter 2
Self-healing strategies in polymer

2.1 Introduction

Mechanical damage such as microcracking, delamination, or fatigue failure is a common degradation pathway in polymeric materials, often leading to a reduction in mechanical strength, barrier properties, and overall performance [1]. Traditional polymer systems lack any intrinsic mechanism to recover from such damage, making them prone to progressive failure over time. In response to this limitation, significant research has been directed toward designing polymers that can autonomously or externally restore their structural integrity through well-defined molecular or structural processes, a class of materials referred to as self-healing polymers [2, 3].

Self-healing in polymers can be achieved through a variety of mechanisms, broadly classified into extrinsic and intrinsic approaches. Extrinsic systems typically involve the incorporation of microcapsules or vascular networks containing healing agents that are released upon damage and subsequently undergo polymerization or solidification to fill the crack. In contrast, intrinsic systems rely on reversible interactions within the polymer matrix itself. These may include reversible covalent bonds (e.g., Diels–Alder adducts, disulfide exchange), dynamic ionic or coordination interactions, or supramolecular motifs such as hydrogen bonding or π–π stacking [3]. These reversible chemistries allow for the reformation of polymer networks when the damaged surfaces are brought into contact under suitable conditions such as heat, pressure, or light exposure.

The efficiency of self-healing depends on multiple factors, including the mobility of the polymer chains, the strength and reversibility of the dynamic interactions, and the time and environmental conditions under which healing occurs. For instance, materials with low glass transition temperatures typically exhibit better segmental mobility at ambient conditions, which facilitates interfacial chain diffusion and network reformation. Conversely, highly cross-linked or glassy polymers may require elevated temperatures or extended timescales for healing to take place due to restricted molecular mobility [1].

In this chapter, we present a comprehensive overview of self-healing strategies in polymers, organized by the type of healing mechanism and the underlying molecular design. We analyze key material parameters that govern healing behavior, such as bond dissociation energies, reaction kinetics, diffusion rates, and rheological properties. Representative systems are discussed with reference to experimental data on mechanical recovery, healing efficiency, and long-term durability. Special attention is given to the thermodynamic and kinetic considerations that determine the extent and rate of healing, as well as the trade-offs between healing ability and other functional properties such as modulus, thermal stability, and processability.

https://doi.org/10.1515/9783111583716-002

2.2 Self-healing strategies in polymer: intrinsic technology

Microcracking would easily be generated in polymers either during manufacturing or in service as a result of mechanical stress or cyclic thermal fatigue. Its propagation and coalescence have to bring about catastrophic failure of the materials and hence significantly deteriorate durability and reliability of the products. Since the tiny damages are mostly difficult to be perceived in time and to repair in particular, the materials had better have the ability of self-healing.

Under inspiration of natural healing in living bodies, healing concepts that offer the ability to restore polymeric materials to their original set of properties have been proposed and successfully applied in recent years [1]. So far, the achievements in this aspect fall into two categories: (i) extrinsic self-healing, which enables healing under certain external stimulation and (ii) intrinsic self-healing. The so-called intrinsic self-healing without manual intervention is only available in a few cases for the time being. As viewed from the predominant molecular mechanisms involved in the healing processes, the reported achievements consist of three modes: (i) physical interactions, (ii) chemical interactions, and (iii) supramolecular interactions. It also operates taking advantage of the intentionally pre-embedded healing agent.

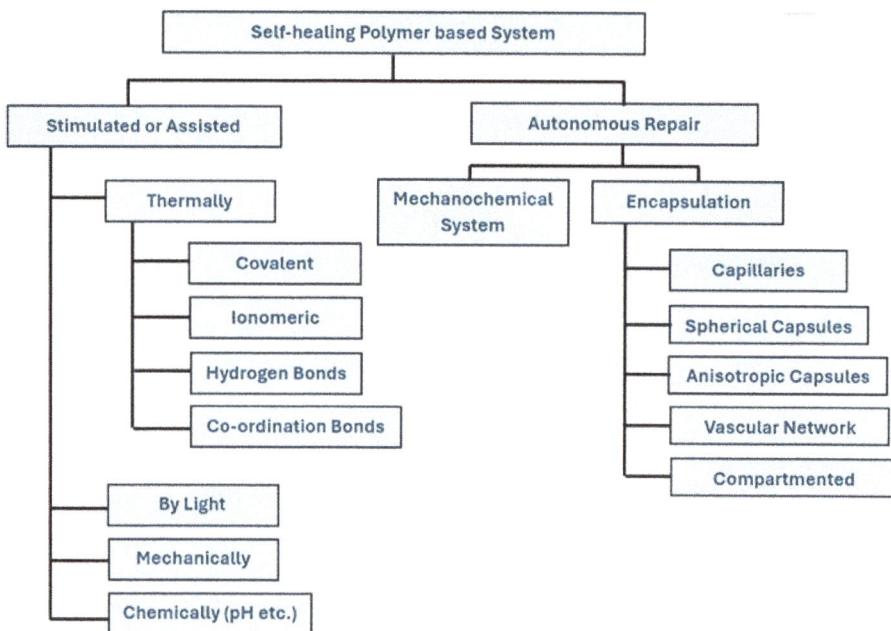

Fig. 2.1: Organization of self-healing polymer-based material systems according to the different principles employed [2].

2.2.1 Release of healing agent

In this case, the healing agent is stored in some media and incorporated into the materials in advance. As soon as the cracks destroy the fragile reservoirs, the healing agent is released into the crack planes due to capillary effect and rebinds the cracks. In accordance with types of the containers, there are two main modes of the repair activity: (i) self-healing in terms of healant-loaded pipelines and (ii) self-healing in terms of healant loaded microcapsules. The crack-triggered delivery of the healing agent allows autonomic rehabilitation without manual intervention.

The core issue of this technique lies in filling the brittle-walled vessels with polymerizable chemicals, which should be fluid at least at the healing temperature. Subsequent polymerization of the chemicals flowing to the damage area plays the role of crack elimination. Dry first identified the potential applicability of hollow glass tubes [3–6]. A similar approach was adopted by Motuku et al. [7] and Zhao et al. [8]. Because the hollow glass capillaries have much larger diameters (on millimeter scale) than those of the reinforcing fibers in composites, they have to act as initiation for composites failure [9]. Instead, Bleay et al. [9] employed hollow glass fiber (with an external diameter of 15 μm and an internal diameter of 5 mm) to minimize the detrimental effect associated with large-diameter fibers. Complete filling of healing agents into the tiny tubes was achieved by vacuum-assisted capillary action filling technique.

Trask et al. [10] considered the placement of self-healing hollow glass fiber layers within both glass fiber/epoxy and carbon fiber/epoxy composite laminates to mitigate damage and restore mechanical strength. The hollow fibers were bespoken with diameters between 30 and 100 μm and a hollowness of approximately 50%. The study revealed that after the laminates were subjected to quasi-static impact damage, a significant fraction of flexural strength can be restored by the self-repairing effect of a healing resin stored within hollow fibers.

Toohey et al. proposed a self-healing system consisting of a three-dimensional microvascular network capable of autonomously repairing repeated damage events [11]. Their work mimicked the architecture of human skin. The 3D microvascular networks were fabricated by deposition of fugitive ink (a mixture of Vaseline/microcrystalline wax (60/40 by weight)) in terms of direct-write assembly [12] through a cylindrical nozzle. Then, the yielded multilayer scaffold was infiltrated with epoxy resin. When the resin was consolidated, a structural matrix was obtained. With the help of heating and light vacuum, the fugitive ink was removed and 3D microvascular networks were created. By inserting a syringe tip into an open channel at one end of the microvascular networks, fluidic polymerizable healing agent was injected into the networks.

As an extension of the hollow glass–fiber approach, vascular networks were also developed by the Bond's group in the University of Bristol for providing a replenishable and repeated self-healing function [13, 14]. They firstly worked out glass fiber/epoxy laminates with vascular sandwich core, giving a relatively thin skinned configuration [13]. The healing networks consisted of tubing bonded into the midplane of the core and vertical risers supplying the skin-core bond region. The distribution of the vertical risers can be varied by design. In another work of them on the same topic [14], a vascular sandwich structure with horizontal supply channels and vertical riser channels in a closed-cell foam core was prepared. Such a network had negligible effect on the baseline static mechanical properties of the composite panel.

With respect to extrinsic self-healing based on capsules, the working principle resembles the aforementioned pipelines but the containers for storing healing agents are replaced by fragile microcapsules. The approach developed by White et al. in the University of Illinois at Urbana-Champaign [15] plays the role of milestone, which is promising for development into a practical technique for mass production and application of the smart materials. They systematically investigated self-healing via ring-opening metathesis polymerization (ROMP) of microencapsulated endo-isomer of DCPD.

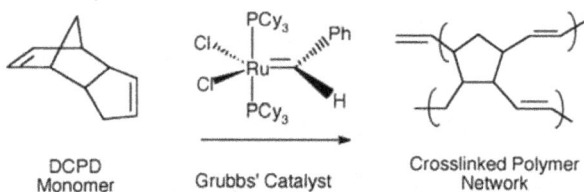

DCPD Monomer   Grubbs' Catalyst   Crosslinked Polymer Network

Fig. 2.2: Ring-opening metathesis polymerization of DCPD.

When damage in the form of a crack ruptured the microcapsules, DCPD was released into the crack plane where it came in contact and mixed with the pre-embedded Grubbs' catalyst (Fig 2.2, 2.3).

As for crack repair in elastomeric systems, Sottos et al. [16] developed a self-healing approach by incorporating two types of polyurethane-formaldehyde-walled microcapsules into a polydimethylsiloxane (PDMS) matrix. These microcapsules were engineered to separately encapsulate the resin and the initiator components required for cross-linking. Specifically, one set of capsules contained a high-molecular-weight vinyl-functionalized PDMS and platinum-based catalyst complexes, while the second set encapsulated a PDMS copolymer with functional groups that could react with the vinyl groups in the presence of the platinum catalyst. Upon mechanical damage, rupture of the microcapsules resulted in the localized release and mixing of these components within the crack plane. The platinum-catalyzed hydrosilylation reaction between the released components enabled in situ polymerization and cross-linking at the damaged interface. A distinguishing characteristic of this system is that the polymer formed during the healing process is chemically identical to the original PDMS matrix, ensuring compatibility and maintaining bulk mechanical properties [16].

In a different approach to self-healing of thermoset polymers, Caruso et al. [17] explored the use of solvents as the active healing agent. Chlorobenzene, a solvent ca-

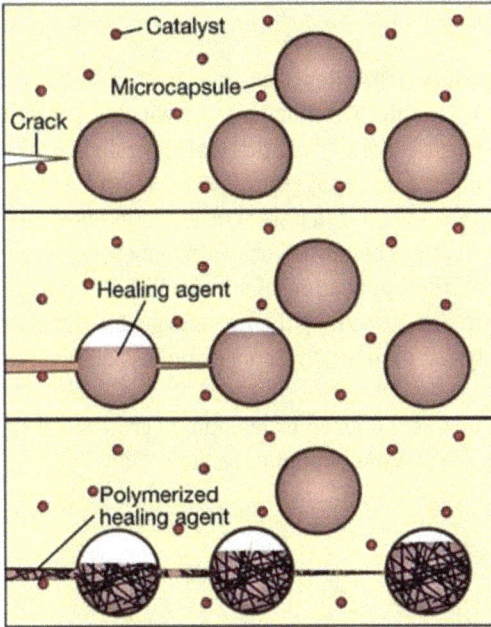

Fig. 2.3: The autonomic healing concept [15]. A microencapsulated healing agent is embedded in a structural composite matrix containing a catalyst capable of polymerizing the healing agent. (a) Cracks form in the matrix wherever damage occurs; (b) the crack ruptures the microcapsules, releasing the healing agent into the crack plane through capillary action; (c) the healing agent contacts the catalyst, triggering polymerization that bonds the crack faces closed.

pable of plasticizing and swelling epoxy networks, was encapsulated within urea-formaldehyde shells and subsequently embedded into an epoxy resin. The underlying principle was that chlorobenzene would locally soften or plasticize the surrounding epoxy matrix, thereby enabling residual reactive sites, left unreacted during incomplete curing, to undergo additional cross-linking and restore network integrity upon solvent exposure. Building upon this concept, Caruso and colleagues further developed a formulation where the core material of the microcapsules was a mixture of an epoxy monomer and chlorobenzene solvent, rather than the solvent alone [18, 19]. This dual-function core allowed the solvent to facilitate molecular mobility while simultaneously delivering fresh monomer to the crack interface. In partially cured epoxy matrices containing residual amine hardeners, the released epoxy monomer could undergo curing reactions at the crack interface, initiated by these residual hardeners. This two-component system improved the interfacial adhesion and resulted in significantly higher healing efficiencies compared to systems employing solvent-only microcapsules, due to the formation of new covalent cross-links directly at the damage site [18, 19].

2.2.2 Reversible crosslink

2.2.2.1 Diels–Alder (DA) and retro Diels–Alder (rDA) Reaction

The Diels–Alder reaction and the retro Diels–Alder reaction have shown a potent pinnacle in fabricating self-healing characteristics, due to their thermal reversibility. In a broad spectrum, the monomer containing the functional groups such as furan or maleimide stacks to form the two carbon–carbon bond in a typical orientation and fabricate the polymer through Diels–Alder reaction. The newly formed polymer, however, breaks down to its original monomer by the application of stimuli like IR or heat, through the retro Diels–Alder mechanism and then the polymer reforms by any suitable condition used to synthesize the polymer earlier [20]. Some examples include the following:

i. A polymer having pendent groups such as the furan or maleimide groups, which then cross-link sequentially through Diels–Alder coupling reaction.

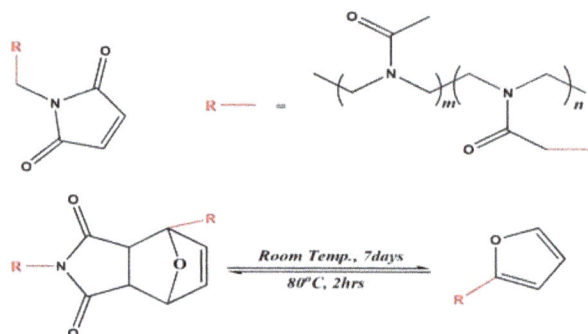

Fig. 2.4: Reversible gelation of polyoxazoline by Diels–Alder [20].

ii. Multifunctional monomers conjugate together to form the polymer using the Diels–Alder pathway. In these reactions, the Diels–Alder reaction occurs in the backbone to fabricate the polymer. Fig. 2.5 shows a polymerization scheme where (3 Ma) tris-maleimide and tetrafuran formed through the Diels–Alder reaction [20].

Another example can be revisited when Zhang et al. [21] fabricated an ultrafast self-healing composites material referenced on Diels–Alder reaction. The surface of the graphene was modified with hydrazine-functionalized graphene oxide and reduction of functionalized graphene oxide was carried out to adopt the hydrazine-functionalized graphene nanosheets. The newly formed composite was synthesized by the incorporation of functionalized graphene nanosheets into pre-polyurethane, which was in turn fabricated from the elimination reaction comprising of NCO terminated polyurethane and a Diels–Alder residue of alcohol and bis-maleimide [22].

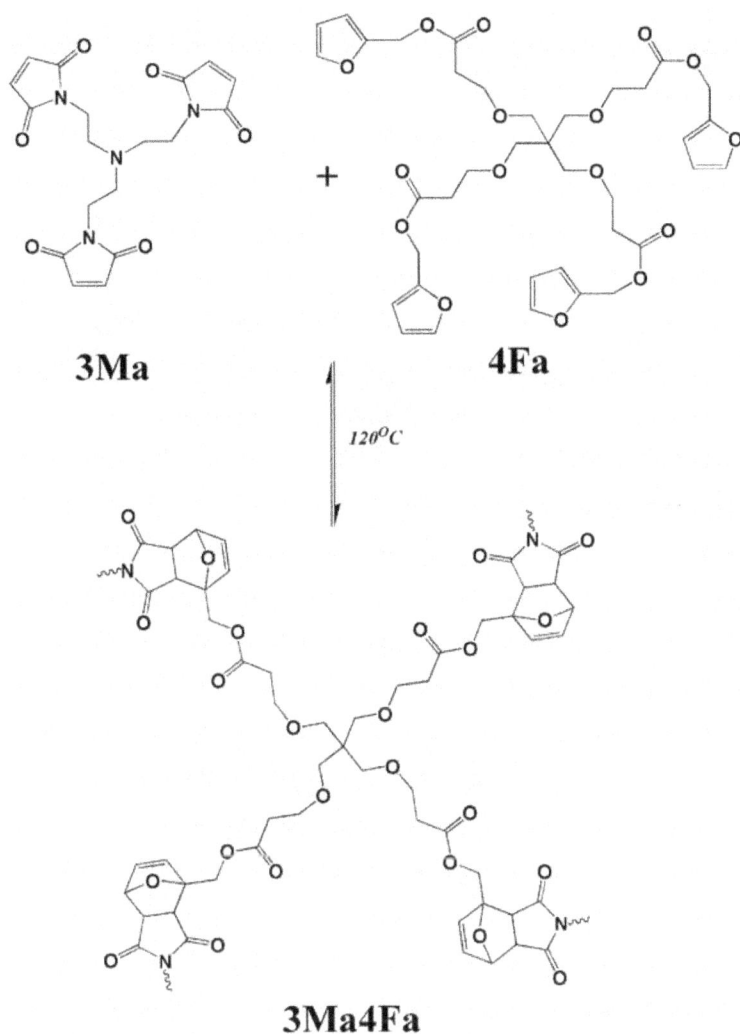

3Ma

+

4Fa

$120^\circ C$

3Ma4Fa

Fig. 2.5: To demonstrate multifunctional monomers conjugation participating in Diels–Alder reaction to induce self-healing [20].

i. Groups participating in Diels–Alder reaction for the self-healing phenomenon

1. Furan and maleimide undergo Diels–Alder reaction under the stimulus of heat keeping several matrices as its building block, for instance, thermosetting epoxy, polyamide, polyurethane, and polyester containing furan. The T_g of the system ranges from 80 °C to 100 °C. Furthermore, keeping the required conditions at 150 °C for 2 h; we can achieve the recovery 1% as high as 72% [23].

Fig. 2.6: Diels–Alder reaction using furan and maleimide (taken from Zeng C, Seino H, Ren J, Hatanaka K, Yoshie N. Bio-based furan polymers with self-healing ability. Macromolecules2013; 46:1794–802).

2. Barthel and his coworkers studied the self-healing effect on thioesters keeping the thermal stimulus as the premise; cyanodithioester and cyclopentadiene also exhibit the shape memory property via the Diels–Alder mechanism. The peo-b-pfge block copolymer or poly (iB$_0$A-nBA) containing cyclopentadiene groups may be used as a material to fabricate the above properties. It has been reported that since the T_g of the PEO-b-PFGE matrix is very low (about −40 °C), the scratch can be recovered immediately stimulating it to 120 °C for 3 h [24].

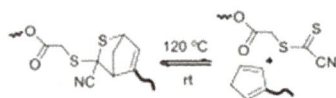

Fig. 2.7: Self-healing effect on thioesters (taken from Barthel MJ, Rudolph T, Teichler A, Paulus RM, Vitz J, HoeppenerS, Hager MD, Schacher FH, Schubert US. Self-healing materials via reversible cross-linking of poly(ethylene oxide)-block-poly(furfuryl glycidyl ether) (PEO-b-PFGE) block copolymer films. Adv Funct Mater 2013;23:4,921–32).

3. Coumarin [2 + 2] cycloaddition takes place where the matrix such as 2, 3-Dicyano-but-2-enedinitrile (TCE) is subjected to a particular wavelength of radiation. The process can yield a recovery rate as high as 72% with a UV radiation at a wavelength greater than 280 nm. Additionally, a thermal potent may be induced in the system to speed up the process [25].

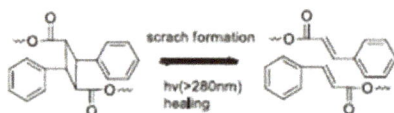

Fig. 2.8: To demonstrate the cinnamoyl [2 + 2] cycloaddition (taken from Chung C-M, Roh Y-S, Cho S-Y, Kim J-G. Crack healing in polymeric materials via photochemical [2 + 2] cycloaddition. Chem Mater 2004;16:3,982–4).

4. Cycloaddition initiates the same way as in the case of coumarin but is generally used in matrixes of thermoplastic polyurethanes of high molecular weight. Short UV wavelength tends to the bonds faster leading to a quick and efficient healing effect [26].

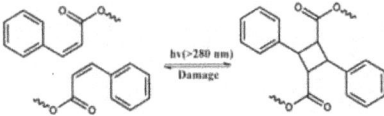

Fig. 2.9: Cycloaddition in coumarin (taken from Ling J, Rong MZ, Zhang MQ. Photo-stimulated self-healing polyurethane containing dihydroxyl coumarin derivatives. Polymer 2012;53:2691–8).

5. In matrixes, comprised of dendritic macromolecules, a special type of cycloaddition reaction based on athracene [4 + 4] cycloaddition takes place. Anthracene having an extended conjugated diene structure can process through the Diels–Alder reaction at a UV wavelength of 254 nm for 15 min. The fracture heals without the loss of any aromaticity yielding a healing efficiency of around 68% [27].

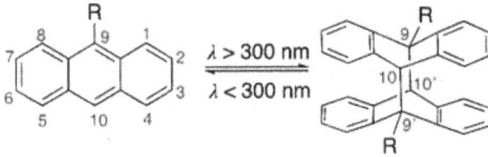

Fig. 2.10: m-cycloaddition reaction of anthracene [4 + 4] (taken from Froimowicz P, Frey H, Landfester K. Towards the generation of self-healing materials by means of a reversible photo-induced approach. Macromol Rapid Commun 2011;32:468–73).

2.2.2.2 Layer-by-layer self-assembly

Layer-by-layer self-assembly technique is one of the most adaptable approaches to process and develop multilayered nonstructural composites. In 1966, Kirkland and Iler began the genesis of practicing this technique by using microparticles [28]. The reason why this process is known to be flexible is because of its recent accomplishments in the preparation of composites by immersion, spin, spray, electromagnetism, or fluidics. A classic example of the illustration of this process can be tracked down by the work of Fan et al. [29] who prepared a self-healing anticorrosion coating on a magnesium alloy (AZ31) surface. The first step comprised of coating cerium nitrate hexahydrate on AZ31 and then heating the specimen for 30 min at 80 °C to convert the Ce (III) to Ce (IV) [29]. Accordingly, poly(ethylene-imine) and graphene oxide were layered on the surface previously achieved by healing to form the PEI/GO layer. Finally, the PEI/GO-coated sample was immersed in PEI, deionized water, and poly (acrylic acid), alternatively to give the surface perfect stability and tangibility. The graphene oxide was incorporated as a corrosion inhibitor and the self-healing ability was a characteristic feature to the PEI/PAA multilayer.

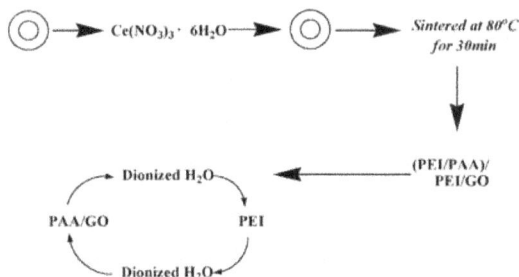

Fig. 2.11: A schematic representation of the fabrication of the PEI/PAA multilayer exhibiting self-healing properties [29].

Sun et al. reported an inherently healable, reduced graphene oxide-bolstered polymer composite via the above-discussed technique [30]. The reduced graphene oxide was modified with β-cyclodextrin, which had the capability to form the complex with branched PEI grafted with ferrocene groups coupled by the host – guest interaction to form Ce PEI-Fc/RGO-CD complexes. These were layer by layer assembled with PAA to give rise to a composite film of polyacrylic acid and Ce PEI-Fc/RGO-CD. The host–guest interaction and their reversible induction between the nanofillers and assembled polyelectrolyte film promote the composite to append excellent mechanical robustness and highly efficient self-healing property [30].

2.2.2.3 Supramolecular assembly

Metal–ligand complex
Polymers that incorporate metal centers into their structure have been reported to exhibit stimulus-responsive self-healing behavior [31]. This capability arises primarily from dynamic metal–ligand coordination bonds that act as reversible, non-covalent cross-links within the polymer network. Unlike covalent bonds, these coordination bonds can dissociate and reassociate under specific external stimuli such as temperature, light, or solvent exposure allowing the network to reorganize, flow into damaged regions, and restore its original mechanical integrity [32]. The metal centers serve not only as reversible bonding sites but also as conformational anchors that, upon activation by an external trigger, can reorganize the polymer chain network to facilitate

healing. The modular tunability of coordination strength and ligand exchange kinetics makes these systems particularly effective for rapid and efficient repair processes.

A notable example of such a system was developed by Williams et al., who synthesized an organometallic conductive polymer incorporating N-heterocyclic carbenes (NHCs) as coordinating ligands [33]. The self-healing functionality was attributed to the dynamic coordination between NHCs and transition metals. Upon exposure to DMSO vapor at 150 °C, fractured samples underwent rapid healing, with the reformed material recovering electrical conductivity on the order of 10^{-3} S/cm. The mechanism was proposed to involve a stimulus-induced dynamic equilibrium between the polymer chains and the metal centers, allowing network segments to flow into the fracture and reform coordination bonds at the interface. This reversible reorganization enabled efficient healing and partial restoration of function without requiring permanent cross-links or external adhesives [33].

Additional studies have demonstrated the versatility of metal–ligand coordination in enabling autonomous or stimulus-induced healing across a range of polymer systems. For instance, Zn(II)–bis(terpyridine) complexes [34] and 2,6-bis(1'-methylbenzimidazolyl) pyridine ligands [77] have been shown to enable healing efficiencies exceeding 75%, due to the dynamic and reversible nature of their coordination with metal ions. These systems are particularly effective due to the high association constants and fast ligand exchange rates associated with these complexes, which facilitate rapid stress relaxation and bond reformation upon damage.

Other transition metal–ligand interactions, such as Cu–O and Si–O coordination bonds, have also been explored for their ability to impart self-healing properties to polymeric materials [35]. In such systems, mechanical damage results in partial cleavage of metal–ligand interactions, which upon exposure to thermal or solvent stimuli, allow for re-coordination and network recovery. Some studies have reported an increase in the elastic modulus in the healed region compared to the pre-damaged state, suggesting densification or improved network rearrangement after repair [36]. These enhancements are often attributed to localized rearrangement of polymer chains and improved interfacial adhesion resulting from dynamic reformation of coordination bonds.

Furthermore, in certain cases, the transition metals involved may form chloride clusters as a consequence of bond cleavage during mechanical stress. These clusters can later redistribute and participate in restoring the network when healing conditions are applied. The reorganization at the molecular scale often results in repaired zones with mechanical and interfacial properties closely matching or even exceeding those of the undamaged material. This reproducible and homogeneous recovery mechanism makes transition metal-based coordination systems a valuable platform for developing self-healing materials capable of restoring both structural and functional properties after damage.

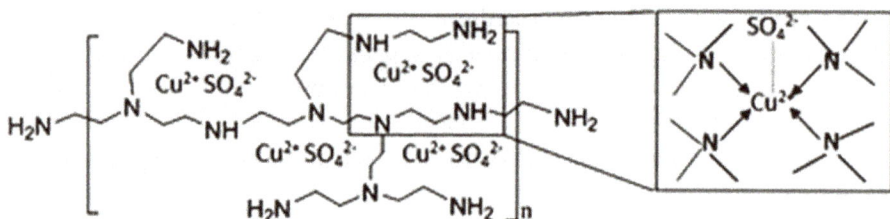

Fig. 2.12: To show metal–ligand bond participating in self-healing pathways (taken from Wang Z, Urban MW. Facile UV-healable polyethylenimine–copper(C2H5N–Cu) supramolecular polymer networks. Polym Chem2013;4:4897–901).

Metathesis

Different from thermal reversible DA reaction, dynamic exchange reactions between reversible covalent bonds offer another opportunity for the development of self-healing polymers. Disulfide bonds can undergo metathesis exchange reactions in which two neighboring S–S bonds are ruptured and reformed through free radical or ionic intermediates.

Seon –Mi Kim et al. reported a transparent and easily processable thermoplastic polyurethane (TPU) having efficient room temperature self-healing ability through facile aromatic disulfide metathesis engineered by hard segment-embedded aromatic disulfides [37]. Hard segments with an asymmetric alicyclic structure provide the optimal metathesis efficiency for the embedded aromatic disulfide while preserving the remarkable mechanical properties of TPU. The recovery of mechanical properties in respliced sample was more than 75% than that of virgin TPU.

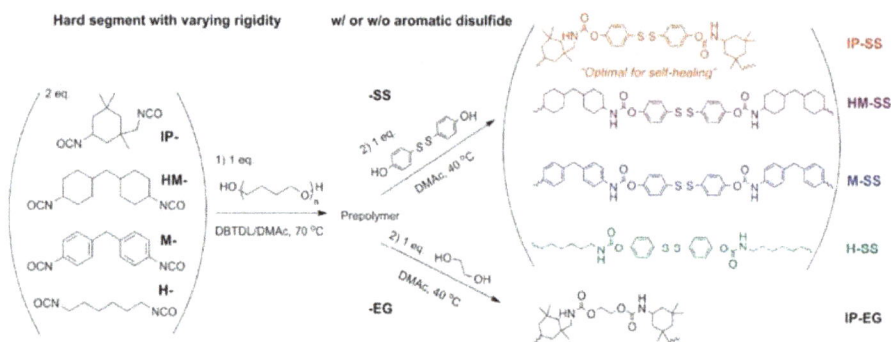

Fig. 2.13: a) Synthetic routes to TPUs with four different diisocyanates (i.e., **IP, HM, M,** and **H**) and two chain extenders (i.e., **SS** and **EG**). The TPUs are designated as X–Y, where X and Y denote the abbreviation of the diisocyanate monomer and chain extender, respectively.

Because of its unique durability and elasticity, material with isophorone diisocyanate (IP)-based hard segment shows promising application in wearable sensors that can repair after routine scratches. Based on the demonstration of a scratch-detecting and auto-repairing electrical sensor coated on IP–SS, Kim and his group suggest that this film has potential applications in the wearable electronics industry.

Yoshifumi Amamoto et al. synthesized a polymer through RAFT copolymerization of n-butyl acrylate (BA) and trithiocarbonate (TTC) cross-linker and demonstrated photo-responsive self-healing owing to reshuffling reaction involving TTC unit [38]. The repeatable photo-responsive

nature of the self-healing system and macroscopic fusion of separate polymer pieces by UV irradiation of the TTC cross-linked polymers were first accomplished in the presence of anisole solvent. Next, the TTC cross-linked polymers were subjected to the self-healing reaction in bulk state, resulting in macroscopic fusion of completely separate polymer pieces. The authors reported successful repetition of the self-healing process at least five times.

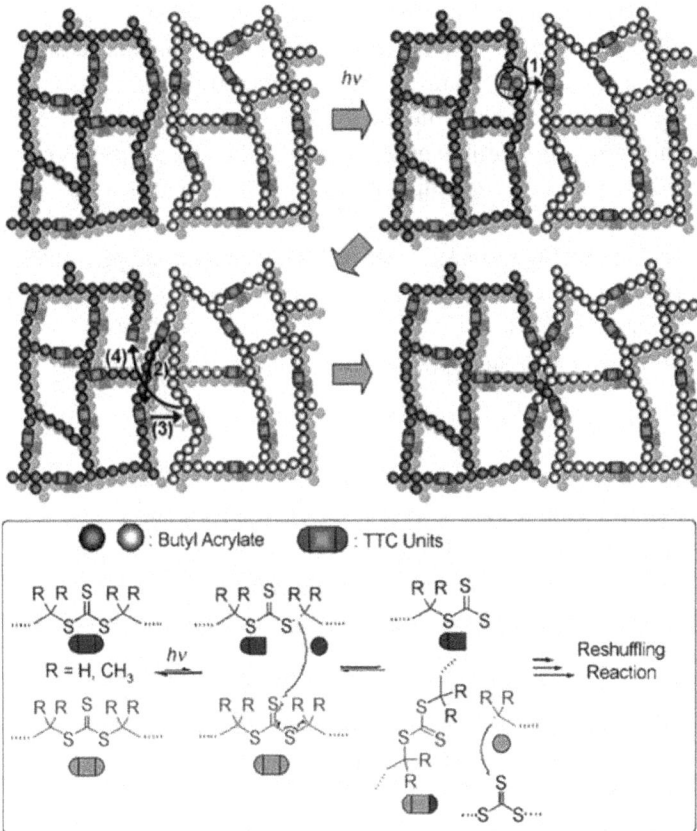

Fig. 2.14: Model self-healing reactions by UV irradiation [38].

H-bonding

Although hydrogen bonds (H-bonds) are weaker than common covalent bonds, due to their directionality and affinity, a wide range of supramolecular polymers with diverse mechanical properties ranging from supramolecular gel to strong rubber like materials can be prepared.

Meijer and coworkers were the first to assemble ureidopyrimidone (Upy) monomers by using quadruple hydrogen bonding non-covalent interactions with high degree of polymerization [39]. The resulting material display mechanical properties similar to traditional polymers. This discovery of using weak reversible hydrogen bonding interactions to produce supramolecular assemblies with high association constant and having polymeric properties makes this field an exciting area for materials research. The Upy compounds are cheap and can be incorporated into other polymeric systems to improve processability or other functionalities.

Cordier and colleagues used diacids and triacids in a two-step synthetic route involving condensation of acid groups with an excess of diethylenetriamine, followed by reactions with urea; a mixture of oligomers containing H-bonding motifs including amidoethyl imidazolidone, di(amidoethyl) urea, and diamido tetraethyltriurea is obtained [40]. Associations of the H-bonding motifs lead to the formation of the self-repairing network. The resulting supramolecular assembly is a translucent glassy plastic, which behaves like soft rubber at temperatures up to 90 °C, and is able to self-repair at ambient conditions. While formation of H-bonds between amide C O and amine-functionalized chains is responsible for network reformation, the presence of ~15% mobile phase enables high healing efficiency at room temperature without any other stimuli. These heterogeneities are usually neglected due to the difficulty of detection, but in fact significantly affect network mobility.

π–π Stacking

π–π Stacking interactions achieved by end-capped π–electron-deficient groups interacting with π–electron-rich aromatic backbone molecules were utilized in the development of thermal-triggered reversible self-healing supramolecular polymer networks. Burattini et al. observed same kind of phenomenon while employing chain-folding copolyimide (electron-deficient) and pyrenyl (electron-rich) end-capped polyamide [41] chains. The Tg of the network can be tuned to achieve self-healing at relatively wide temperature range (~50–100 °C) by changing the spacer and adjusting the composition of the blend. Upon heating, the π–π stacking interactions will be interrupted, enabling pyrenyl end-capped chains to disengage from copolyimide and flow due to the presence of a flexible "soft" spacer. Thus, repair of damage and regaining of mechanical strength by reformation of the π–π stacking will occur.

Xu et al. reported a novel polymer system based on nitrobenzoxadiazole (NBD)-functionalized cholesterol (Chol) derivatives that exhibited a distinctive combination of π–π stacking and intermolecular hydrogen bonding interactions [42]. In this sys-

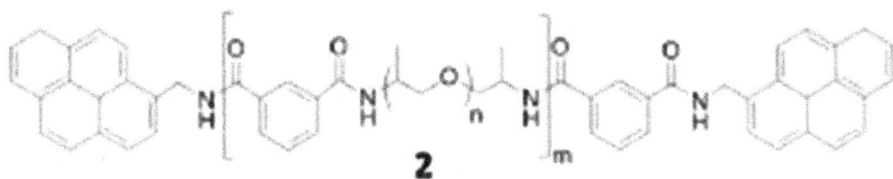

Fig. 2.15: Polyamide end-capped with pyrenyl groups [41].

tem, the molecular architecture was carefully designed such that the NBD moiety, known for its aromatic electron-deficient character, and the cholesterol unit, a rigid and hydrophobic steroidal group, were linked through a flexible spacer. The length of this spacer played a pivotal role in dictating the overall self-assembly behavior of the molecules. Specifically, the π–π stacking interactions arose primarily from the planar, conjugated NBD groups, which can stack in a face-to-face manner due to favorable electron cloud overlap. This stacking promotes the formation of ordered, extended supramolecular structures critical for gelation. Concurrently, intermolecular hydrogen bonding among functional groups on the cholesterol derivatives contributed to additional non-covalent cross-linking that reinforced the three-dimensional network within the gel matrix.

By varying the length of the spacer between the NBD and Chol units, Xu et al. demonstrated control over the balance and strength of these two non-covalent interactions. A shorter spacer favored tighter packing and stronger π–π interactions, resulting in gels with higher mechanical stability but potentially slower self-healing due to restricted molecular mobility. Conversely, longer spacers increased chain flexibility, allowing for more dynamic hydrogen bonding and easier reorganization of the network, thereby enhancing the healing efficiency of the gel.

This molecular design strategy showed that subtle modifications at the molecular level could fine-tune the gelation and self-healing properties of supramolecular materials. Such control over intermolecular interactions is critical for developing soft materials that combine mechanical robustness with the ability to autonomously repair after damage, which has significant implications for applications in biomedicine, soft robotics, and responsive materials.

Ionic interaction
Ionic interactions in polymers play a significant role in enabling self-healing behavior, particularly through the formation of ionomers – polymers that contain a small fraction of ionizable groups covalently attached to the polymer backbone. These ionic moieties can interact to form reversible cross-links or clusters that respond dynamically to mechanical stress or thermal stimuli. Among such systems, poly(ethylene-co-methacrylic acid) (EMAA) is a well-documented example that exhibits self-healing

properties following mechanical damage. In projectile puncture tests, EMAA demonstrated the ability to undergo polymer chain diffusion into the damaged zone under both ambient and elevated temperature conditions, thereby facilitating partial or complete healing of the fractured area [43].

The proposed mechanism for healing in EMAA-based ionomers is understood to operate via two interconnected processes [44]. Upon mechanical impact, the propagation of a crack within the polymer matrix disrupts the ionomeric network. This disruption, coupled with local deformation and friction, generates heat in the vicinity of the crack tip. The thermal energy not only promotes the mobility of polymer chains but also leads to local softening or melting of the matrix, which is crucial for subsequent healing. Following this initial phase, the healing proceeds through chain diffusion across the fracture interface. During this stage, individual polymer chains move into the damaged region, leading to reconnection and partial restoration of the polymer network. Given sufficient time and thermal energy, the chains undergo relaxation and reorganization, ultimately resulting in the reformation of a mechanically cohesive structure across the previously damaged interface [45].

The efficiency of this healing process is influenced by several critical parameters, including the rate of chain diffusion, the size and mobility of ionic clusters within the polymer, and the heating rate achieved during or after damage. Chain diffusion is strongly dependent on temperature; as the system approaches the melting point of the ionomeric matrix, chain mobility increases significantly, enhancing the probability of interfacial entanglement and healing [46]. However, the presence of large ionic clusters can hinder this process. Larger cluster sizes tend to restrict the motion of individual chains, effectively immobilizing parts of the network and reducing the polymer's ability to flow or reconfigure during healing. In addition, these clusters increase the thermal inertia of the material, thereby lowering the effective heating rate in the impacted region [47].

Interestingly, it has been reported that certain stimuli such as thermal or mechanical energy can disrupt these ionic clusters. The dissociation of clusters under these conditions increases local chain mobility and accelerates healing, even in the absence of added chemical agents [48]. This effect has been demonstrated in several ionomer systems, where autonomously triggered dissociation of ionic aggregates led to measurable improvements in healing kinetics and mechanical recovery [48].

Beyond EMAA, other ionically cross-linked polymer systems have also demonstrated promising self-healing capabilities. For example, polymer gels derived from the complexation of polyamines with multivalent phosphate-bearing anions were shown to form physically cross-linked networks capable of robust self-repair. These gels exhibited excellent underwater adhesion to both hydrophilic and hydrophobic surfaces, attributable to the strong reversible ionic interactions between the oppositely charged components. Upon mechanical disruption, the ionic cross-links could reversibly dissociate and reform, enabling the material to autonomously restore its mechanical integrity under suitable conditions [49].

Host–guest inclusion

Host–guest interactions refer to non-covalent binding phenomena wherein one molecular entity (the "host") forms a supramolecular complex with another (the "guest") through spatial and chemical complementarity. When polymer segments with distinct structural motifs overlap in a two-dimensional arrangement and engage through weak, reversible forces – despite their chemical dissimilarity – such interactions can be classified under this category [50]. These interactions are inspired by a wide range of natural molecular recognition events, such as enzyme–substrate binding or protein–ligand docking, where specificity and reversibility coexist. Leveraging this principle, recent developments in polymer science have enabled the design of synthetic host–guest systems that are not only reversible under physiological or environmental stimuli but also endow the material with self-healing capabilities [51].

One well-characterized example involves the reversible complexation between cyclodextrin, a cyclic oligosaccharide with a hydrophobic inner cavity, and ferrocene, a redox-active organometallic compound. In particular, polyacrylic acid chains terminated with ferrocene groups can act as guests, while cyclodextrin units grafted onto another polymer backbone serve as hosts. The ferrocene-cyclodextrin host–guest complex forms spontaneously through hydrophobic interactions between the ferrocene moiety and the interior of the cyclodextrin cavity. Upon the application of a suitable external stimulus – such as a change in redox potential, temperature, or pH – a sol–gel transition is triggered, leading to reversible cross-linking and network formation. This process facilitates macroscopic self-healing, as the network can disassemble and reassemble without permanent damage to the polymer structure [52].

These host–guest interactions are often stabilized by non-covalent forces such as π–π stacking, van der Waals interactions, and hydrophobic effects. Due to their dynamic, reversible nature, these interactions are particularly suitable for developing stimuli-responsive hydrogels with regenerative mechanical properties. Beyond self-healing, their utility extends into various biomedical applications, including targeted drug delivery, tissue scaffolding, and responsive biosensors. In drug delivery systems, for example, the selective binding of drug molecules (guests) to host polymers can be used to control the release rate or trigger site-specific delivery based on external cues, thereby improving therapeutic efficiency and minimizing side effects.

2.3 Stimuli responsive self-healing strategies on polymer: extrinsic technology

2.3.1 Heating

For a surplus number of self-healing processes, thermal initiation is one of the major sources to proceed to the healing process. This can be phased in with the Diels–Alder reaction chemistry where a reaction takes place in between a diene and a dienophile.

A good pair of a diene and a dienophile can start the healing process at room temperature itself. Murphy et al. fabricated a series of self-healing Diels–Alder-based monomer that uses the cyclopentadiene as both the diene and the dienophile [53]. A cyclic derivative of DCPD, which in term serves as the Diels–Alder adduct between the two molecules of cyclopentadiene was developed using several snips of organic linkers in such a way that the rDA reaction of the cycle would yield the desired polymers. It was inferred that to retard back to its original dimensions, graphene/polymer composite need to repair themselves intrinsically, which requires certain external stimuli such as mechanical force, light, heat, or pH changes.

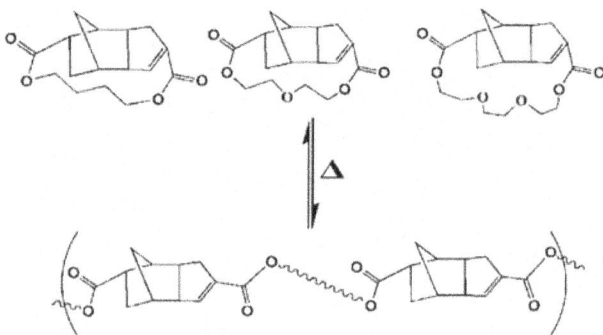

Fig. 2.16: Assisted self-healing mechanism [53].

Heating is widely recognized as an effective and practical method for enhancing self-healing in polymer composites due to its relative simplicity and low operational cost. The application of heat increases the mobility of polymer chains within the material, allowing them to diffuse and rearrange across damaged interfaces. This increased molecular mobility facilitates the reformation of physical or chemical bonds, thereby improving the self-healing efficiency and mechanical recovery of the composite and related materials [54].

In one notable study, Pungo et al. developed a composite material consisting of silicone rubber (SR) integrated with graphene nanoparticles (GNPs) that exhibited a negative temperature coefficient behavior, meaning its electrical resistance decreased as temperature increased. This composite demonstrated significant self-healing capability when subjected to a thermal annealing process at 250 °C for two hours [55]. Post-annealing, the tensile strength of the SR/GNP composite improved by approximately 87%, indicating substantial mechanical recovery. The underlying mechanism was attributed to reversible cross-linking within the damaged polymer network, particularly involving free silicon groups that serve as reactive sites. Upon heating, these groups facilitated the reassembly of the fractured network through thermally activated bonding, allowing the composite to restore both its structural integrity and functional properties.

Conductive polymer composites, however, often suffer a decrease in electrical conductivity with repeated mechanical loading due to the disruption of conductive pathways formed by conductive fillers such as graphene. With increasing load cycles, microcracks and polymer chain displacements sever these percolation networks, resulting in diminished conductivity. Contrasting this common limitation, Zhan et al. fabricated a conductive composite comprising graphene dispersed within a thermoplastic polyurethane matrix, which exhibited remarkable recovery of electrical properties after thermal treatment [56]. Specifically, after four cycles of tensile loading followed by thermal annealing, the electrical conductivity of the composite was found to nearly double, compared to its original value. This observation implies that the thermal treatment not only repaired mechanical damage but also facilitated the reformation of conductive networks disrupted by mechanical strain. The heat-induced polymer chain mobility and the potential reorganization of graphene fillers during annealing allowed the conductive pathways to reconnect, thereby restoring and even enhancing electrical conductivity [57].

These studies collectively demonstrate that thermal stimuli serve as an efficient external trigger for activating molecular and nanoscale reconfigurations in polymer composites, enabling them to recover mechanical strength and electrical performance after damage. This approach is particularly advantageous in applications where materials are exposed to repetitive mechanical stress and require durable self-healing to extend their functional lifespan.

2.3.2 Microwave

An element having large area of conjugated π structure is able to absorb microwaves, which makes it a protean element in the field of self-healing composites. For instance, the π structure of graphene, under microwave absorption, will make it a huge electric dipole and subsequently transform microwaves in the form of dipole distortion, which eventually results in the heating. The ability of graphene to absorb microwaves due to the large area of conjugated π structure makes it a protean element in the field of self-healing composites [58]. Zhang et al. used the Diels–Alder synthesis route to prepare covalently cross-linked reduced functionalized GO/PU composites with self-healing property using microwave [59]. The two broken surfaces were autonomously healed when subjected to gentle pressure and exposed to an 800-W microwave oven operating in the frequency range of 2.45 GHz for 5 min and then allowed to heat at 70 °C for 2 h without any external continuous pressure. The microwaves absorbed by reduced functionalized graphene oxide turned into heat and then accelerated the healing process of the composite. Karak et al. developed a reinforced self-healing nanocomposite based on hyperbranched polyurethane containing iron oxide nanoparticle decorated reduced graphene oxide (IORGO), which demonstrated healing efficiency of more than 99% using microwave process fabrication [60].

2.3.3 Solvent-assisted self-healing

A pragmatic external stimulus to impart the self-healing properties is deionized water or organic solvent [61]. Solvents play a pivotal role in the reuniting of chemical bonds. Zhang et al. had carried out an experiment with the copolymerization of thermoplastic polyurethane and diacetyl carbonate using graphene oxide as a compatibilizer, which exhibited the self-healing properties in presence of dropping water [62]. The fragmented pieces were relatively put together with the broken and fractured surfaces in contact with each other. Because of the adhesiveness of the newly formed fractured surface from the hydrogel, no superlative forces were required for connecting the broken parts. The drop of water healed the sample yielding tensile stress of 503.4 kPa and the healing efficiency is 92.3% [62]. The results confirm and infer that ionic bonds and hydrogen bonds can be reformed via water assistance.

Adding another dimension to the above process, salt solution can also help to ease the self-healing process. Wand and Tang et al. had fabricated a multiple shape memory, super tough and self-healable PAA-GO-Fe^{3+} hydrogel [63] in which the healing process was facilitated by the ionic binding of Fe^{3+} ions to the carboxyl groups on the PAA chains. The self-healing process was introduced by immersing the fractured surface in $FeCl_3$/HCl solution for about 15 h.

The hydrogel, when exposed to the atmospheric condition, can bear a dumbbell of 5.5 kg. Interestingly the fractured position of the healed hydrogel after the tensile test is not the healing position, concluding the novel standardization of the cut surfaces by Fe^{3+} ions.

2.3.4 Electromagnetic radiation

UV or IR light radiation can also be a potent stimulus in the specification of self-healing polymeric composite. Cho et al. identified one of the mechanisms by the [2 + 2] cycloaddition of cinnamoyl groups. Tricinnamates were fabricated and radiated via photons to form cyclobutane containing cross-linked polymer films. Several studies were conducted for the same; it was found that under UV light stimulations, the cyclobutyl groups at the fractured regions reverted themselves to reform the cinnamoyl functionalities [64]. Scientists provided a two-step hypothesis for the light radiation healing mechanisms. Firstly the damage had surfeited the pendant groups generating two reactive ends [64]. The exposure to the UV rejoined the fractured polymer matrix region in the backbone, which eventually combined with the reactive ends of the pendant groups. Both close and seal scratch damage resulted in high efficiency in their healing rate. One of the most superior inferences from this process lies in the fact of rapid healing ($\approx$ 30 min) and even micron level cracks can be healed by the UV wavelength, which is similar to that of the sunlight.

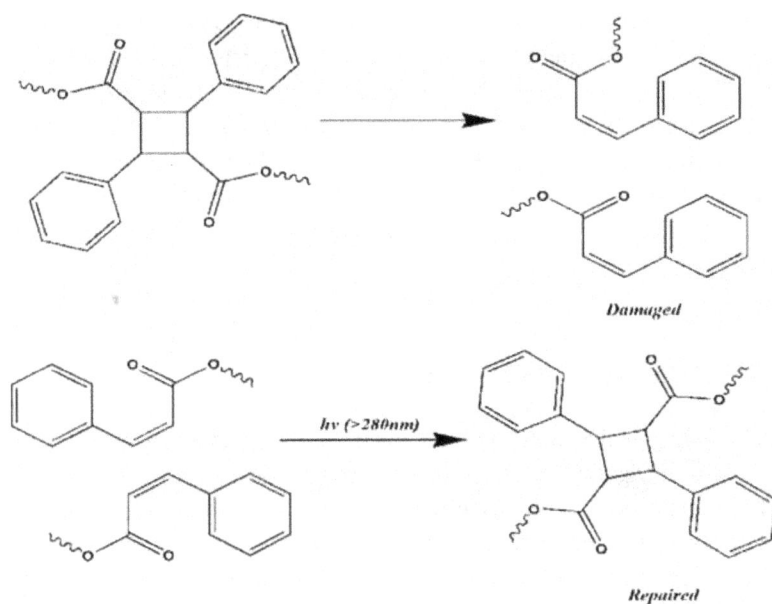

Fig. 2.17: UV-based crack healing in cyclobutane containing polymers of cinnamoyl groups [64].

The near-infrared irradiation (NIR) has been patented as a highly efficient skin-penetrating biomedical technique. Tong et al. fabricated fast self-healing graphene oxide hectorite clay – poly (*N, N* – dimethyl acrylamide) hybrid hydrogel supplemented by NIR irradiation for 2–3 mins. The process resulted in the hydrogel to achieve a strong recovery of 96% [65]. Not only did graphene oxide act as a complemented cross-linking agent, but also as an NIR irradiation absorber to convert it to thermal energy simultaneously with efficiently promoting both-way diffusion of the polymer chain across the surfaces. Liang et al., they developed a self-healing bilayer hydrogel system by conducting an experiment. The healing of the fracture surfaces was achieved by irradiating with an NIR photo excitement having a wavelength of 808 nm and a power of 1.25 W. The self-healing characteristics of the developed polymer system were attributed to the photothermal energy transformation property of graphene oxide. It was also reported that with the increase of the graphene oxide content in the system, the heating rate of the hydrogel increased [66].

Kim and his coworkers developed a composite comprised of polyurethanes and graphene nanocomposites, which demonstrated improved mechanical, thermal, and optical property [67]. The self-healing ability of polyurethane/graphene nanocomposite was elucidated because of the intermolecular diffusion of polymer chains, which can be fastened up by NIR absorption.

Apart from IR and NIR, there are several other wavelengths of light radiation that can result in the self-healing capability. Fei et al. developed a tri-layer, light-triggered healable and highly electrically conductive fibrous membrane by imposing reduced graphene oxide and silver nanowires onto gold nanoparticles layered poly (ε-caprolactone) [68]. Under 532-nm light irradiation, the polymer chains inter diffuse amongst them to heal the cracked surface of the damaged fibrous membrane, and re-crystallize upon cooling when the wave diode laser was switched off. Not only did the surface conductivity recovered by 91%, but the tensile strength of the membrane along with the elongation at break was also still maintained excellently after multiple cutting–healing cycles.

2.3.5 Simple contacting: autonomous healing

Autonomous self-healing, which occurs without the need for any external stimuli, represents a particularly intriguing area of research because it enables the healing process to take place at ambient conditions, often at room temperature. This capability eliminates the necessity for additional energy input such as heat, light, or chemical triggers, making these materials highly practical for everyday applications where spontaneous repair is desired. The mechanism of autonomous self-healing in polymers typically involves a sequence of three fundamental steps.

The first step is triggering or actuation. When the polymer experiences damage, such as a chain scission or fracture, the molecular alignment or orientation within the material is disturbed. This disruption acts as a signal to initiate the self-healing process by activating the mobility or reactivity of the polymer chains near the damaged region. Essentially, the broken chains or interfaces become sites of increased molecular activity, which begins the repair process.

The second step involves the molecular interactions that facilitate the healing itself. Polymers containing functional groups such as hydroxyl (-OH), siloxane (Si–O–Si), and amine (-NH$_2$) are especially conducive to autonomous healing because these groups can form reversible non-covalent bonds like hydrogen bonds or π–π stacking interactions. These interactions serve as temporary "bridges" between fractured surfaces, enabling the material to reconnect without permanent chemical changes. An important factor influencing this step is the polymer's degree of amorphousness and its glass transition temperature. Polymers with lower Tg values have chains that are more flexible and mobile at room temperature, allowing polymer segments to diffuse more readily across the damaged interface. This increased chain mobility accelerates the formation of hydrogen bonds, covalent bonds, or π–π interactions, thereby facilitating rapid restoration of the fracture surface.

The final step is the healing response itself, which refers to how quickly the material can restore its structural integrity. Since autonomous self-healing does not rely on any external triggers or catalysts, the intrinsic properties of the polymer, such as chain mobility and bonding kinetics, govern the rate of healing. This response time is a critical parameter because it determines the practical effectiveness of the material for real-world applications. The balance between polymer chain flexibility, density of functional groups capable of reversible bonding, and the polymer's physical state all contribute to how efficiently and rapidly the material can recover from damage under ambient conditions.

Keeping the economic and operational aspect in mind, recent advances have been made to achieve the self-healing attributes in the polymer without any external stimuli [69]. Peng and Turung et al. had crafted a mussel-inspired electroactive chitosan/graphene oxide composite, which demonstrated the self-healing properties just by bringing in contact the two fractured surfaces [70]. The above property was exhibited due to the dynamic covalent bonds, hydrogen bonding, and π–π structure. While plotting the stress–strain curve, Peng and his coworkers concluded that the recovered composite had similar mechanical properties as the initial one. Poly-borosiloxane is a well known "solid–liquid" material whose viscoelastic properties promote fast and complete healing due to the presence of dative bonds, which can be activated in dynamic stressed condition [71]. The mechanical and the electrical properties can be healed with the healing recovery of 98% just by keeping the fractured surfaces in contact for about 10 min. Since the healing is promoted by polymer flow and dynamic dative bond interaction between the polymer chains, the healing process can be continued for more than five cycles [71].

Self-healing speed is another determining factor from the pragmatic view of the material. A high-speeded self-healing material is always preferred due to the increased cycles of loading in a fixed time span. Bao et al. fabricated self-healing thermal reversible elastomer matrixes on amine-terminated randomly branched oligomer and graphene oxide [72]. The self-healing properties can be achieved without using any plasticizer, solvent, healing agent, or external stimuli. The amorphous chain morphology of the polymer coupled with its low glass transition temperature allows the polymer to diffuse and coordinate amongst them at room temperature. The two fractured surfaces were rejoined in less than 60 s. The composite (with 1% graphene oxide content) was reported to heal up to 60% of its original tensile strength preceding the fracture and bringing them back in contact for 1 min. Increasing the amount of graphene oxide gives room for more covalent cross-linking, which in turn results in more restricted movement of the polymeric chains, and hence the time taken for the healing process is elevated. A second experiment was carried out increasing the graphene oxide content to 2% and 4%. The samples too resulted in a 36% and 20% recovery from their original extensibilities, respectively, in 1 min. The self-healing capability only dropped to 90% efficiency as compared to the original after the two cut surfaces were left apart for 24 h and sloped down to 50% after 96 h of delay [72].

2.4 Conclusion

Thus, self-healing strategies in polymers rely on the ability of polymer chains to restore broken bonds and recover material properties after damage. Heating is a widely used method because it increases chain mobility, allowing polymer segments to move and re-form connections at the damaged site. This process is effective in composites where dynamic cross-links or reversible bonds exist, enabling the material to regain mechanical strength and, in some cases, electrical conductivity.

Autonomous self-healing polymers operate at room temperature without external input. Their healing relies on functional groups such as hydroxyl, amine, or siloxane that form reversible hydrogen bonds or π–π interactions. The mobility of polymer chains, governed by factors such as the glass transition temperature and the degree of amorphous regions, determines how quickly and efficiently the material can self-repair. Lower glass transition temperatures facilitate faster chain diffusion and quicker healing.

Additionally, metal–ligand coordination and host–guest interactions provide further mechanisms for self-healing. These reversible bonds allow the material to reorganize at the molecular level and restore the fractured network. Polymers incorporating these interactions have demonstrated high healing efficiencies and can recover both structural and functional properties.

The effectiveness of self-healing depends on the balance between polymer chain mobility, the nature and density of reversible bonding sites, and environmental conditions. Understanding these parameters at the molecular level is crucial for designing polymers with predictable healing behavior and sufficient mechanical recovery. Continued investigation into these mechanisms will support the development of durable polymers for applications where material longevity and damage tolerance are essential.

References

[1]   Zhang, M. Q. & Rong, M. Z. (2011). Self-healing Polymers and Polymer Composites. John Wiley & Sons, Inc: Hoboken.
[2]   Fischer, H. (2010). Self-repairing material systems—a dream or a reality? *Natural Science*, *2*, 873–901.
[3]   Dry, C. (1992). Passive tunable fibers and matrices. *International Journal of Modern Physics, B, 6*, 2763–2771.
[4]   Dry, C. (1994). Matrix cracking repair and filling using active and passivemodes for smart timed release of chemicals from fibers into cementmatrices. *Smart Materials and Structures*, *3*, 118–123.
[5]   Dry, C. & McMillan, W. (1996). Three-part methylmethacrylate adhesive systemas an internal delivery system for smart responsive concrete. *Smart Materials and Structures*, *5*, 297–300.
[6]   Dry, C. (1996). Procedures developed for self-repair of polymer matrixcomposite materials. *Composite Structures*, *35*, 263–269.

[7]   Motuku, M., Janowski, C. M. & Vaidya, U. K. (1999). Parametric studies onself-repairing approaches for resin infused composites subjected tolow velocity impact. *Smart Materials and Structures, 8*, 623–638.

[8]   Zhao, X. P., Zhou, B. L., Luo, C. R., Wang, J. H. & Liu, J. W. (1996). A model of intelligentmaterial with self-repair function (in Chinese). *Chinese Journal of Materials Research, 10*, 101–104.

[9]   Bleay, S. M., Loader, C. B., Hawyes, V. J., Humberstone, L. & Curtis, P. T. (2001). Asmart repair system for polymer matrix composites. *Composites Part A: Applied Science and Manufacturing, 32*, 1767–1776.

[10] Trask, R. S., Williams, G. J. & Bond, I. P. (2007). Bioinspired self-healing of advancedcomposite structures using hollow glass fibres. *Journal of the Royal Society Interface, 4*, 363–371.

[11] Toohey, K. S., Sottos, N. R., Lewis, J. A. & Moore, J. S. (2007). White SR.Self-healing materials with microvascular networks. *Nature Materials, 6*, 581–585.

[12] Therriault, D., Shepherd, R. F., White, S. R. & Lewis, J. A. (2005). Fugitive inks fordirect-write assembly of three-dimensional microvascular networks. *Advances in Materials, 17*, 394–399.

[13] Williams, H. R., Trask, R. S. & Bond, I. P. (2008). Self-healing sandwich panels:Restoration of compressive strength after impact. *Composites Science & Technology, 68*, 3171–3177.

[14] Williams, H. R., Trask, R. S. & Bond, I. P. (2007). Self-healing composite sandwichstructures. *Smart Materials and Structures, 16*, 1198–1207.

[15] White, S. R., Sottos, N. R., Geubelle, P. H., Moore, J. S., Kessler, M. R., Sriram, S. R., Brown, E. N. & Viswanathan, S. (2001). Autonomic healing of polymercomposites. *Nature, 409*, 794–797.

[16] Keller, M. K., White, S. R. & Sottos, N. R. (2007). A self-healing poly(dimethylsiloxane) elastomer. *Advanced Functional Materials, 17*, 2399–2404.

[17] Caruso, M. M., Delafuente, D. A., Ho, V., Moore, J. S. & Sottos, N. R. (2007). White SR.Solvent-promoted self-healing materials. *Macromolecules, 40*, 8830–8832.

[18] Caruso, M. M., Blaiszik, B. J., White, S. R., Sottos, N. R. & Moore, J. S. (2008). Fullrecovery of fracture toughness using a nontoxic solvent-based self-healing system. *Advanced Functional Materials, 18*, 1898–1904.

[19] Blaiszik, B. J., Caruso, M. M., McIlroy, D. A., Moore, J. S., White, S. R. & Sottos, N. R. (2009). Microcapsules filled with reactive solutions for self-healingmaterials. *Polymer, 50*, 990–997.

[20] Wu, S., Li, J., Zhang, G., Yao, Y., Li, G., Sun, R. & Wong, C. (2017). Ultrafast Self-Healing Nanocomposites via Infrared Laser and Their Application in Flexible Electronics. *ACS Applied Materials and Interfaces, 9*(3), 3040–3049. doi: 10.1021/acsami.6b15476.

[21] Sin, J., Zhang, W. & Li, Y. (2017). Reduced graphene oxide reinforced polymeric film with excellent mechanical robustness. *ACS Nanotechnology, 11*, 7134–7141.

[22] Hou, C., Huang, T., Wang, H., Yu, H., Zhang, Q. & Li, Y. (2013). A strong and stretchable self-healing film with self-activated pressure sensitivity for potential artificial skin applications. *Scientific reports, 3*, 3138. doi: 10.1038/srep03138.

[23] Brown, E. N., White, S. R. & Sottos, N. R. (2005). Retardation and repair of fatigue cracks in a microcapsule toughened epoxy composite – Part II: In situ self-healing. *Composites Science & Technology, 65*, 2474–2480.

[24] Barthel, M. J., Rudolph, T., Teichler, A., Paulus, R. M., Vitz, J., Hoeppener, S., Hager, M. D., Schacher, F. H. & Schubert, U. S. (2013). Self-Healing Materials via Reversible Crosslinking of Poly(ethylene oxide)-Block-Poly(furfuryl glycidyl ether) (PEO-b-PFGE) Block Copolymer Films. *Advanced Functional Materials, 23*, 4921–4932. doi: 10.1002/adfm.201300469.

[25] Brown, E. N., Sottos, N. R. & White, S. R. (2002). Fracture testing of a self-healing polymer composite. *Experimental Mechanics, 42*, 372–379.

[26] Zhang, L., Wang, W. & Dan, Y. (2017). Self-healing Cellulose Membranes Prepared by Microcapsules Containing UV-initiated Healing Agents. *DEStech Transactions on Materials Science and Engineering*. doi: 10.12783/dtmse/amsee2017/14249.

[27] Shenmei, W., Yuan, L., Aijuan, G., Zhang, Y. & Liang, G. (2016). Synthesis and characterization of novel epoxy resins-filled microcapsules with organic/inorganic hybrid shell for the self-healing of high performance resins: Novel Epoxy Resins-Filled Microcapsules with Hybrid Shell. *Polymers for Advanced Technologies, 27.* doi: 10.1002/pat.3829.

[28] Kirkland, J. J. (1965). Porous Thin-Layer Modified Glass Bead Supports for Gas Liquid Chromatography. *Analytical Chemistry, 37*(12), 1458–1461. doi: 10.1021/ac60231a004.

[29] Fan, F., Zhou, C., Wang, X. & Szpunar, J. (2015). Layer-by-layer assembly of a self-healing anticorrosion coating on magnesium alloys. *ACS Applied Materials & Interfaces, 7,* 27271–27278.

[30] Xiang, Z., Zhang, L., Li, Y., Yuan, T., Zhang, W. & Sun, J. (2017). Reduced graphene oxide -reinforced polymeric films with excellent mechanical robustness and rapid and highly efficient healing properties. *ACS Nanotechnology, 11,* 7134–7141.

[31] Mozhdehi, D., Ayala, S., Cromwell, O. R. & Guan, Z. (2014). Self-healing multiphase polymers via dynamic metal-ligand interactions. *Journal of the American Chemical Society, 136,* 16128–16131.

[32] Rao, Y. L., Chortos, A., Pfattner, R., Lissel, F., Chiu, Y. C., Feig, V., Xu, J., Kurosawa, T., Gu, X. & Wang, C. (2016). Stretchable self-healing polymeric dielectrics cross-linked through metal-ligand coordination. *Journal of the American Chemical Society, 138,* 6020–6027.

[33] Chen, Y., Kushner, A. M., Williams, G. A. & Guan, Z. (2012). Multiphase design of autonomic self-healing thermoplastic elastomers. *Nature Chemistry, 4,* 467–472.

[34] Shi, Y., Wang, M., Ma, C., Wang, Y., Li, X. & Yu, G. (2015). A Conductive self-healing hybrid gel enabled by metal-ligand supramolecule and nanostructured conductive polymer. *Nano Letters, 15,* 6276–6281.

[35] Novoselov, K. S., Geim, A. K., Morozov, S. V., Jiang, D., Zhang, Y., Dubonos, S. V., Grigorieva, I. V. & Firsov, A. A. (2004). Electric field effect in atomically thin carbon films. *Science, 306,* 666–669.

[36] Zhu, J., Yang, D., Yin, Z., Yan, Q. & Zhang, H. (2014). Graphene and graphene-based materials for energy storage applications. *Small, 10,* 3480–3498.

[37] Kim, S.-M., Jeon, H., Shin, S.-H., Park, S.-A., Jegal, J., Yeon Hwang, S., Oh, D. X. & Park, J. (**2017**). Superior Toughness and Fast Self-Healing at Room Temperature Engineered by Transparent Elastomers. *Advances in Materials,* 1705145.

[38] Amamoto, Y., Kamada, J., Otsuka, H., Takahara, A. & Matyjaszewski, K. (2011). Repea*table* Photoinduced Self-Healing of Covalently Cross-Linked Polymers through Reshuffling of Trithiocarbonate Units. *Angewandte Chemie International Edition, 50,* 1660–1663.

[39] Beijer, F. H., Sijbesma, R. P., Kooijman, H., Spek, A. L. & Meijer, E. W. (1998). *Journal of the American Chemical Society, 120,* 6761.

[40] Cordier, P., Tournilhac, F., Soulié-Ziakovic, C. & Leibler, L. (2008). Self-healing and thermoreversible rubber from supramolecular assembly. *Nature, 451,* 977–980.

[41] Burattini, S., Colquhoun, H. M., Fox, J. D., Friedmann, D., Greenland, B. W., Harris, P. J. F., Hayes, W., Mackay, M. E. & Rowan, S. J. (2009). A self-repairing,supramolecular polymer system: healability as a consequence of donor–acceptor π–π stacking interactions. *Chemical Communication, 44,* 6717–6719.

[42] Xu, Z., Peng, J., Yan, N., Yu, H., Zhang, S., Liu, K. & Fang, Y. (2013). Simple design but marvelous performances: molecular gels of superior strength and self-healing properties. *Soft Matter, 9,* 1091–1099.

[43] Kalista, S. J. & Stephen, J. (2003). Self-healing of thermoplastic poly (ethylene-co-methacrylic acid) copolymers following projectile puncture. Virginia Polytechnic Institute and State University: Blacksburg, VA, 67. M.S. Thesis.

[44] Mather, P. T., Qin, H., Wu, J. & Bobiak, J. (2006). POSS-based polyurethanes: from degradable polymers to hydrogels. In: Medical Polymers 2006 (International Conference Focusing on Polymers used in the Medical Industry, 5th, Cologne, Germany, RAPRA, Shrewsbury, UK, pp. 5/1–5/9.

[45] Adachi, H., Yokoi, T., Hatori, T., Morishita, K., Sakashita, K., Kaiya, H., Inoue, K., Ueda, Y., Nakamura, T. & Yamaguchi, S. (1990). Temperature Display Devices. Jpn. Pat., 02124438.

[46] Lin, C., Lee, S. & Liu, K. (1990). Methanol-Induced crack healing in poly (methyl methacrylate). *Polymer Engineering & Science, 30*, 1399–1406.

[47] Beloshenko, V. A., Varyukhin, V. N. & Voznyak, Y. V. (2005). The shape memory effect in polymers. *Russian chemical reviews, 74*, 265–283.

[48] Prager, S. & Tirrell, M. (1981). The healing process at polymer–polymer interfaces. *Journal of Chemical Physics, 75*, 5194–5198.

[49] Huang, Y., Lawrence, P. G. & Lapitsky, Y. (2014). Self-assembly of stiff, Adhe-sive and self-healing gels from common polyelectrolytes. *Langmuir, 30*, 7771–7777.

[50] Zhang, C., Ren, Z., Yin, Z., Qian, H. & Ma, D. (2008). Amide II and amide III bands in polyurethane model soft and hard segments. *Polymer Bulletin, 60*, 97–101.

[51] Doddi, S., Ramakrishna, B., Venkatesh, Y. & Bangal, P. R. (2015). Photo-driven near-IR fluorescence switch: synthesis and spectroscopic investigation of squarine-spiropyrandyad. *RSC Advances, 5*, 97681–97689.

[52] Klajn, R. Spiropyran-based dynamic materials. Chemical Society Reviews.

[53] Zhu, P., Hu, M., Deng, Y. & Wang, C. (2016). One-Pot Fabrication of a Novel agar-polyacrylamide/graphene oxide nanocomposite double network hydrogel with high mechanical properties. *Advanced Engineering Materials, 18*, 1799–1807.

[54] Cui, W., Ji, J., Cai, F. Y., Li, H. & Rong, R. (2015). Robust, anti-fatigue, and self-healing graphene oxide/ hydrophobic association composite hydrogels and their use as recyclable adsorbents for dye wastewater treatment. *Journal of Materials Chemistry A, 3*, 17445–17458.

[55] Valentini, L., Bon, S. B. & Pugno, N. M. (2016). Severe graphene nanoplatelets aggregation as building block for the preparation of negative temperature coefficient and healable silicone rubber composites. *Composites Science and Technology, 134*, 125–131.

[56] Yuan, C., Rong, M. Z., Zhang, M. Q., Zhang, Z. P. & Yuan, Y. C. (2011). Self-Healing of Polymers via Synchronous Covalent Bond Fission/Radical Recombination. *Chemistry of Materials, 23*, 5076–5081.

[57] Li, J., Zhang, G., Deng, L., Zhao, S., Gao, Y., Jiang, K., Sun, R. & Wong, C. (2014). In situ polymerization of mechanically reinforced, thermally healable graphene oxide/polyurethane composites based on Diels-Alder chemistry. *Journal of Materials Chemistry A, 2*, 20642–20649.

[58] Hu, H., Zhao, Z., Wan, W., Gogotsi, Y. & Qiu, J. (2013). Ultralight and highly compressible graphene aerogels. *Advances in Materials, 25*, 2219–2223.

[59] Zhu, J. J., Kirkland, Y., Murali, S., Stoller, M. D., Velamakanni, A., Piner, R. D. & Ruoff, R. S. (2010). Microwave assisted exfoliation and reduction of graphite oxide for ultracapacitors. *Carbon, 48*, 2118–2122.

[60] Thakur, S. & Karak, N. (2015). A tough smart elastomeric bio-based hyperbranched polyurethane nanocomposite. *New Journal of Chemistry, 39*, 2146–2154.

[61] Han, D. & Yan, L. (2013). Supramolecular hydrogel of chitosan in presence of graphene oxide nanosheets as 2D cross linkers. *ACS Sustainable Chemistry & Engineering, 2*, 296–300.

[62] Zhang, Q., Chen, Q. & Liu, L. (2016). Tough Stretchable, compressive novel polymer/graphene oxide nano composites double network hydrogel with high mechanical properties. *Advanced Engineering Materials, 18*, 1799–1807.

[63] Tang, W. & Wang, T. (2016). Multiple shape memory; self-healable; and super tough PAA-GO-Fe3+ hydrogel, Macromol. *Materials Engineering, 302*.

[64] Wu, M. & Sun, J. (2014). Rapid and efficient multiple healing of flexible conductive films by near infrared irradiation, ACS. *Applied Materials & Interfaces, 6*, 16409–16415.

[65] Tong, Z. & liu, W. (2014). Fast self-healing of graphene oxide heatorite clay poly (N,N –dimethyl acrylamide) hybrid hydrogels realized by near infrared radiation. *ACS Applied Materials Interfaces, 6*, 22855–22861.

[66] Kim, E., Jin, K. & Kim, B. (2013). Synthesis and properties of near IR induced self-healable PU/ graphene nanocomposites. *European Polymer Journal, 49*, 3889–3896.

[67] Sijbesma, R. P., Beijer, F. H., Brunsveld, L., Folmer, B. J., Hirschberg, J. K., Lange, R. F., Lowe, J. K. & Meijer, E. (1997). Reversible polymers formed from self-complementary monomers using quadruple hydrogen bonding. *Science, 278,* 1601–1604.

[68] Fei, B., Wu, F. & Chen, L. (2016). Electrical and mechanical self-healing membrane using gold nano particles as localized 'nano heater. *The Journal of Materials Chemistry C, 4,* 10018–10025.

[69] Peng, R. & Yang, Y. (2014). Conductive nanocomposite hydrogels with self healing property. *RSC Advances, 4,* 35149–35155.

[70] Peng, X. & Turung, L. S. (2017). Mussul Inspired electroactive chitosan/ graphene oxide hydrol with rapid self-healing behavior for tissue engineering. Carbon, Vol. 125, pp. 557–570.

[71] Liu, N., Aller, R. & Bao, Z. (2013). A rapid & efficient self-healing thermos-reversible elastomer crosslinked with graphene oxide. *Advances in Materials, 25,* 5785–5790.

[72] Kin, J. & Park, H. S. (2015). Self healable graphene based composites with sensing capabilities. *Advances in Materials, 27,* 4788–4794.

Chapter 3
Fabrication method of self-healing polymer compounds

3.1 Introduction

The fabrication of self-healing polymer compounds is a critical aspect that determines their structural integrity, healing efficiency, and practical applicability. This chapter focuses on the various methods used to prepare polymer systems that possess intrinsic or extrinsic self-repair capabilities. The fabrication process involves the careful selection and incorporation of dynamic bonding motifs such as reversible covalent bonds, hydrogen bonds, metal–ligand coordination, or host–guest interactions into the polymer matrix. Additionally, the processing conditions, including temperature, mixing techniques, and curing protocols, play a significant role in defining the microstructure and distribution of healing agents within the material.

Different fabrication approaches such as bulk polymerization, solution casting, emulsion polymerization, and microencapsulation are discussed in detail. The integration of microcapsules or vascular networks containing healing agents is one widely explored strategy for extrinsic self-healing polymers, while intrinsic self-healing systems rely on dynamic bonds distributed throughout the polymer network. Each fabrication method influences the polymer's morphology, mechanical properties, and the accessibility of the healing functionalities.

Therefore, understanding the relationship between fabrication parameters and the resulting polymer architecture is essential to optimize self-healing performance. This chapter provides an in-depth analysis of the processing techniques used to create self-healing polymers, highlighting their advantages, limitations, and potential applications. Such knowledge is foundational for advancing the design and manufacture of polymers with enhanced durability and autonomous repair capabilities.

3.2 Melt mixing

Melt blending is one of the most widely employed methods for preparing polymer composites containing graphene, largely due to its cost-effectiveness and environmental friendliness. The process typically occurs within single or twin-screw extruders, where the polymer matrix and nanoparticles are simultaneously melted and mixed [1]. Within the extruder, shear and tensile forces are applied, which serve to break down nanoparticle agglomerates, thereby improving the dispersion and uniformity of graphene within the polymer matrix. Although extruders offer versatility in processing, allowing for modifications such as screw configuration changes to enhance dis-

https://doi.org/10.1515/9783111583716-003

persion, studies have shown that combining an internal mixer with extrusion generally produces more homogeneous dispersions and stronger filler-polymer interactions. This is attributed to the higher shear rates achievable in internal mixers, which facilitate better exfoliation and distribution of the filler, provided processing parameters are optimized [2, 3].

A key advantage of melt blending is that it is a solvent-free process, which enhances its suitability for industrial applications due to reduced environmental impact and simplified processing. The compounded composite material produced through melt blending can subsequently be shaped by secondary processing methods such as injection molding, profile extrusion, or blow molding. However, optimizing melt blending parameters remains a significant challenge due to the many variables involved. Factors such as temperature, screw speed, residence time, and applied shear stress all interact to influence the quality of dispersion. Minor deviations in these parameters can lead to the formation of agglomerates, which alter packing density and reduce the aspect ratio and purity of the filler, ultimately compromising the uniformity of the composite [3].

In some cases, the shear forces generated during processing may be insufficient to fully exfoliate or disperse graphene sheets within the polymer matrix. While specialized mixers capable of producing higher shear rates can improve dispersion in these cases, they also increase production costs. Moreover, excessively high shear forces can generate significant heat within the polymer melt, potentially leading to thermal degradation of the polymer matrix [4]. This thermal sensitivity limits the efficiency of melt blending for polymers with low thermal stability.

Another consideration is the inherently low bulk density of graphene, which complicates uniform dispersion during melt blending. Despite these challenges, successful fabrication of graphene–polymer composites has been demonstrated with various polymers, including polylactic acid (PLA), polyethylene terephthalate (PET), polypropylene (PP), Nylon 6, polycarbonate (PC), and polystyrene (PS). However, high shear rates can induce physical defects in graphene sheets – such as buckling, rolling, or shortening – which diminish their reinforcing effectiveness and reduce filler activity [5–10].

Melt blending is also extensively applied to the fabrication of polymer composites containing nanoclay fillers. Polymers such as polystyrene (PS), polyolefins, polycarbonate, polycaprolactone (PCL), lactides, polymethyl methacrylate (PMMA), polyamides, acrylonitrile butadiene styrene (ABS), and polyethylene oxide (PEO) have been successfully compounded with nanoclays, achieving varying degrees of exfoliation and enhanced mechanical properties [11, 12]. Microwave irradiation has been used as an adjunct technique to improve intercalation and exfoliation during the melt blending of clay-PEO composites.

For example, Lie et al. prepared exfoliated composites using Nylon 6 as the polymer matrix and organically modified clay as the filler in a twin-screw extruder. Due to the inherent low affinity between the polymer and the silicate surface of the clay, mechanical stresses generated during melt blending were essential to delaminate agglomerates and facilitate dispersion [13]. The degree of filler dispersion was directly correlated with the polymer-clay affinity; stronger interactions resulted in more homogeneous composites. Since polyolefins lack polarity, they are often chemically modified with polar groups, such as maleic anhydride, to improve compatibility with clay fillers. This modification enhances the interfacial adhesion and leads to improved mechanical properties compared to unmodified polyolefins [14, 15].

Furthermore, longer residence times in the extruder increase the duration of applied shear, promoting better dispersion of clay particles within the polymer. Increased melt viscosity also contributes to higher shear stresses, which further facilitates the exfoliation and uniform distribution of clay platelets throughout the matrix [16].

3.3 Solution mixing

Apart from melt mixing, solution mixing or solution casting is another popular method often used to fabricate graphene/polymer composites [17]. The method is based upon the principle of dissolution and evaporation, where the nanoparticles are agitated within a polymer system, which in turn is dissolved in a specific solvent before the final casting process and evaporation. The process is equally efficient with using thermoplastic or thermosetting polymers yielding comparable tensile properties along with toughness. Till date, several polymers like PMMA, poly (vinyl alcohol), polyhydroxyaminoether (PHAE), PS, PE, PEO, and epoxy are blended with graphene and other nanofillers such as carbon tubules [18–20].

The process is advantageous as compared to the melt mixing in the sense that the reduced viscosity in case of a polymer in a solvent as compared to that of the melt, together with agitation (mechanical stirrer or ultrasonication) results in the better dispersion of the filler platelets in the polymeric sub network [21].

However, the presence of solvents may be a threat to the environment and removal of these solvents may use certain sophisticated technologies like membrane separation and solvothermal process, which can influence the cost of the final output [22].

Kuilla and his team had shown that the solution-blending technique can be very useful in preparing composites, which have various issues in the dispersion of the filler phase within the polymer matrix [23]. They had prepared a number of modified graphene /polymer composites using polymer matrices as PVA, PMMA, PP, PS, LLDPE, nylon, epoxy, PANI, and PU [24, 25].

The usage of casting process has also been studied with clay as nanofillers and using organic solvents in polymer matrices like PEO, lactide, and PVA. As soon as the solvent is added, the clay particles get swollen in the solvent. Sequentially, the polymer solution is added to suspend the clay during the process, which is coupled with heating and dispersion. The polymer chains further intercalate and substitute the solvent within the layers between the two stacks of clay [26]. It is to be noted that the amount of intercalation of the polymer chains into the clay platelets is highly dependent upon the polarity of the solvent. In fact, the polarity of the solvent and the solvent–polymer interaction is the dominating factor, which influences the mechanical properties of the composite in the solution-blending process [27].

There have been works, which have tried to synthesize the polymer graphene composites for several water-soluble polymers such as PEO, PVA, and poly (vinyl pyrrolidone) (PVP), using solvents such as toluene, chloroform and, acetonitrile [28]. It has been reported that the percentage of the graphene dispersion within the matrix coupled with the volume fraction of the filler drastically decreases [29, 30].

Researchers have recently shifted to another well-known technique for the synthesis of nanocomposites and to overcome the above problems in solution mixing known as in situ polymerization. The process involves heterophasic polymerization reaction in aqueous medium making the process both economical and environmental friendly [32].

3.4 In situ polymerization

"In situ" is derived from its Latin counterpart, which means "in place." As the name suggests, in this particular process, the monomer and the graphene oxide were mixed prior to the addition of initiator and its subsequent polymerization [33]. Green et al., in their experiment, had synthesized physically cross-linked graphene polyacrylamide self-healing hydrogel with a superior quality of thermal and electrical conductivity with the help of this preparation route. All the reactants, acrylamide, N, N-methylbisacrylamide (MBA), and potassium persulfate were mixed together and stirred with the graphene dispersion in water, resulting in the polymerization [34]. An interesting fact to note here is that in "in situ" polymerization process, graphene can be further modified and participates in the polymerization process, giving rise to the formation of covalent bond between graphene and the polymer that is being used [35]. Karak et al. devised a tough self-healing elastomeric nanocomposite, which involved the blends of castor oil-based hyperbranched polyurethane and iron oxide nanoparticles-surfaced graphene oxide nanohybrid [35]. The IORGO was fabricated by the coprecipitation of ferrous and ferric ions on the graphene oxide layers, which was succeeded by the reduction of graphene oxide by using sodium borohydride. The entire process of the manufacturing was carried out "in situ" with poly (ε-caprolactone) diol, 1, 4-butanediol, 2, 4-toluene diisocyanates, and IORGO [36]. This resulted in the formation of a prepolymer. A monoglyceride of castor oil as a chain extender was introduced to form the final nanocomposite of PU/IORGO. The propagation reaction process was dominantly a condensation reaction but traces of anionic side reaction may be found when the conversion rate crosses 60% [37]. In a similar way, reduced graphene oxide coated with sulfur was also prepared, which was followed by the concoction of self-healing hyperbranched PU/SRGO nanocomposite [38].

In situ polymerization can be useful in the sense that it has a better yield along with the advantage of a single-step reaction at a time. Kim et al. had developed a phenyl isocyanate-modified GO, and luminated the self-healing composites by the condensation reaction of poly tetramethylene glycol and 4, 4-methylene diphenyl diisocyanate, and phenyl isocyanate-modified GO in presence of phenylhydrazine [39].

The process of in situ polymerization was first discovered in the late 1980s after which it was made viable for several industrial and biomedical applications. The reason behind its growth and success was perhaps because of the several advantages it possessed [40] – the primary superiority being in the fact that this process supported controllable microcapsule size and thickness coupled with ease of scaling up and low cost. The only disadvantage that it demonstrates is as compared to other encapsulation processes, it takes much more time to complete its healing process [41].

A typical in situ polymerization reaction is characterized by vigorous agitation or sonication of a biphasic liquid such as oil emulsions in water. Generally, the core chemical acts as the dispersed phase while the peripheral solution serves as the continuous phase [42]. The monomers along with the initiators, which are used for fabricating the capsule wall, are made to be dissolved either in the continuous or the dispersed phase. Since the polymer synthesized is often insoluble in the emulsion, polymerization often initiates on the surface of the core material droplets [43]. At times, it may also happen that the newly developed polymer settles onto the droplet surface, generating microcapsules with the core material.

It is a trend that most capsules, which are employed in the process of self-healing, are synthesized by in situ or interfacial polymerization in an oil–water emulsion system to complete the polymerization process. The first self-healing microcapsule was developed by polymerizing poly (Urea-Formaldehyde) containing DCPD as the wall material [44]. The obtained microcapsules were spherical in shape having an average diameter in the rage of 10-1,000 microns and had controlled rheological parameter after been left for drying [44, 45]. The shell thickness was consistently ranging from 160 nm and 220 nm [44–46]. As time progressed, this particular experiment turned out to be the reference for scientists working in the field of self-healing using in situ polymerization techniques. For instance, Yuan et al. prepared PUF capsules with epoxy monomers [47] while Suryanarayana et al. developed linseed oil-based PUF microcapsules [48]. Gragert et al. synthesized an advanced encapsulation technique, which comprised of the reaction between an azide and an alkyne via in situ polymerizations [49].

Poly (urea-formaldehyde) was gradually replaced by poly (melamine-formaldehyde) to achieve better weather-resistant property together with good sealing characteristics and optimized brittleness. Yuan et al. fabricated poly (melamine-formaldehyde) capsules with diglycidyl tetrahydro-o-phthalate and polyphenol [50, 51]. The former was a low viscosity and highly active epoxy monomer while the later was a low-temperature hardener for epoxy materials. At 250 °C for 24 h, these materials showed brilliant efficacies in self-healing against fatigue tests [50, 51].

As technology evolved, these materials were replaced by styrene phased PMF microcapsules and GMA phased PMF microcapsules, especially to attribute the self-healing behavior in thermoplastic (PMMA and PS) and thermosetting polymers (cured epoxy) [52].

Since PMF was a costlier product as compared to PUF, copolymerization strategy for these materials was the best-optimized solution to obtain a consolidated property keeping the cost within limits [53]. A comprehensive polymer system was developed by Liu et al. using poly (melamine-urea-formaldehyde) microcapsules having ENB-based self-healing agents [54]. The catalyst used for this system was Grubbs' Catalyst, which was effective both in terms of economy and time consumption. The PMUF microcapsules demonstrated a better overall property as compared to their UF self-healing composites. The methods flourished since the fabrication method of PMUF microcapsules was much simpler as compared to UF.

Further scientific developments enabled technologists to use a high-temperature-resistant wall material. Heat-resistant polyphenylene oxide (PPO) microcapsules containing glycidyl ether of bisphenol A epoxy were synthesized via in situ polymerization [55]. The process was fabricated developing PPO resin via oxidative polymerization of 2, 6-dimethyl phenol at 30–50 °C in presence of a copper amine complex catalyst (Cu-ethylenediamine). The microcapsules remained stable even at 258 °C proving the viability of using the same in high-temperature environment [55].

In the double capsules healing system, the healing agent, which is polymerized along with the curing agent are separately encapsulated, which turns out to be a problematic approach in terms of pragmatic viewpoint [55]. Firstly, a homogeneous dispersion comprising of two separate types of microcapsules is very difficult to achieve. To ensure the properly adjusted contact in between the healing components, double-layer microcapsules are prepared for encapsulation of both epoxy monomer and the hardener (diamino-diphenyl sulfone) into individual microcapsules [56]. The synthesis process follows a complex process where MF polymerizes to form epoxy-dispersed microcapsules via in situ polymerization, sequenced by the physical adsorption of DDS on the surface of the MF microcapsule via electrostatic interactions or electro ionic interactions [57]. The performance of these specific self-healing substances remains questionable since it is not possible to achieve the self-healing attribute at room temperature. The answer to the probation lies in the fact that the reaction between epoxy and DDS is not feasible under a room-temperature stimulus.

The preparation of the microcapsules by in situ polymerization methods not only depends on the materials used to prepare the core and shell, but also on the reaction conditions, for instance, the type of emulsifier, rotor speed of agitation, core/shell material ratio, pH value, and reaction temperature. Among these, the type of the emulsifier is the most important controlling parameter for the reaction to take place, and this must be selected according to the type of the core material [58]. To cite an example, sodium dodecyl benzene sulfonate (SDBS) should be complemented with polyvinyl alcohol as an emulsifier system for the preparation of the DCPD core microcapsule with a PMF shell [59]. The initial characterization showed a high thermal resistance and stability. But, advanced research revisited the fact that the emulsifier was not suitable because of the high viscosity and small size of the core material [60]. To counterfeit the problem, poly (styrene-malefic sodium) solution was employed as an emulsifier system for encapsulating epoxy DTP. The process yielded small microcapsules with a thin wall, which supplemented the inclusion of the material into its maximum level.

Researches show that the size of the microcapsules can be easily fine-tuned as required by altering the emulsifier content and dispersion rate. However, it is seen that excessively small and large numbers of capsules might turn out to be redundant for the self-healing system. Hence it is always desirable to achieve broad distribution of the capsular agent [61].

Liu et al. had fabricated a Shirasu porous glass (SPG) membrane emulsification process for a customized designed microcapsule of ENB by in situ polymerization [61]. Further, the particle analysis showed that the capsules had a mean diameter of 40 micrometers and a shell thickness of 400–600 nanometer. The advantages this procedure had over the others were that SPG-based microcapsules had a narrower size distribution, which reinforced it to give rise to a stable core. Moreover, the system exhibited thermal stability up to 300 °C. On the other hand, the procedure is very difficult to scale up since it is very complicated and cost-ineffective [61].

3.5 Simple mixing

One of the most traditional and widely used techniques for fabricating graphene or graphene derivative-based polymer composites is simple physical mixing [62]. In this approach, graphene or its oxidized form, graphene oxide (GO), is typically dispersed into a polymer matrix through mechanical stirring, often assisted by ultrasonication. The use of ultrasonication plays a critical role in de-agglomerating graphene nanosheets, promoting uniform distribution within the polymer matrix. For instance, Salezi et al. [63, 64] reported the successful preparation of a ternary composite comprising polyvinyl alcohol (PVA), agar, and graphene, utilizing mechanical stirring under ultrasonicated conditions. This ensured better dispersion of the graphene component throughout the polymer network, thereby improving the uniformity and performance of the resulting material.

However, due to the intrinsic tendency of graphene to aggregate – driven by strong π–π stacking interactions and van der Waals forces between adjacent graphene layers – direct mixing without modification often results in poor dispersion. To overcome this challenge, graphene is frequently oxidized to graphene oxide. The introduction of oxygen-containing functional groups increases the hydrophilicity and colloidal stability of the graphene sheets in aqueous media, thereby facilitating better dispersion in polar polymers [65].

An illustrative example of this is the synthesis of a supramolecular hydrogel system involving chitosan and graphene oxide by Yan et al. [66]. In their study, the gelation behavior was highly dependent on the concentration of graphene oxide. At elevated concentrations of GO, the hydrogel network was observed to form readily at ambient temperature (approximately 27 °C), whereas at lower concentrations of GO, gelation required heating to a much higher temperature (around 95 °C). This observation underscores the critical role of GO concentration in establishing sufficient physical cross-linking via hydrogen bonding and electrostatic interactions with chitosan chains, which together promote the formation of a stable gel matrix. The functional groups present on GO surfaces, such as hydroxyl, epoxy, and carboxyl groups, are primarily responsible for these interactions, further enabling the formation of three-dimensional networks in the composite hydrogel.

3.6 Conclusions

The fabrication of self-healing polymer compounds requires precise control over material composition, processing conditions, and filler dispersion to achieve effective autonomous or stimulus-responsive healing. Among various methods, melt blending remains one of the most scalable and solvent-free approaches, particularly for thermoplastic matrices reinforced with nanofillers like graphene or clay. However, the success of this method depends critically on optimizing shear forces, screw configuration, temperature, and residence time to prevent agglomeration and preserve filler integrity. On the other hand, solution mixing techniques, although limited by solvent use and environmental concerns, offer better control over dispersion at the molecular level, especially for systems involving functionalized graphene or supramolecular networks. The interactions between polymer matrices and fillers, whether covalent, ionic, or hydrogen-bond-based play a pivotal role in ensuring uniform network formation and recovery under damage. Advances in processing, such as high-shear extrusion and sonication-assisted dispersion, continue to refine these methods, allowing the development of materials with tunable mechanical, thermal, and self-healing properties. Future research must focus on integrating these methods with scalable and environmentally sustainable practices while maintaining control over filler–polymer interfacial interactions to maximize healing efficiency.

References

[1]   Grady, B. P. (2011). Carbon Nanotube-Polymer Composites Manufacture, Properties, and Applications. John Wiley and Sons: New York, NY, USA; 145.
[2]   Chen, B., Evans, J. R. G., Greenwell, H. C., Boulet, P., Coveney, P. V., Bowdenf, A. A. & Whiting, A. (2008). A critical appraisal of polymer-clay nanocomposites. *Chemical Society Reviews, 37,* 568–594.

[3] Faucheu, J., Gauthier, C., Chazeau, L., Cavaillé, J. Y., Mellon, V. & Lami, E. B. (2010). Miniemulsion polymerization for synthesis of structured clay/polymer nanocomposites: Short review and recent advances. *Polymer, 51*, 6–17.

[4] Pinnavaia, T. J. (1983). Intercalated Clay Catalyst. *Science, 220*, 365–371.

[5] Coleman, J. N., Khan, U., Blau, W. J. & Gun'ko, Y. K. (2006). Small but strong: A review of the mechanical properties of carbon nanotube-polymer composites. *Carbon, 44*, 1624–1652.

[6] Coleman, J. N., Khan, U. & Gun'ko, Y. K. (2006). Mechanical Reinforcement of Polymers Using Carbon Nanotubes. *Advances in Materials, 18*, 637–640.

[7] McClory, C., Chin, S. J. & McNally, T. (2009). Polymer/Carbon Nanotube Composites. *Australian Journal of Chemistry, 62*, 762–785.

[8] Andrews, R. & Weisenberger, M. C. (2004). Carbon nanotube polymer composites. *Current Opinion in Solid State & Materials Science, 8*, 31–37.

[9] Coleman, J. N., Cadek, M., Blake, R., Nicolosi, V., Ryan, K. P., Belton, C., Fonseca, A., Nagy, J. B., Gun'ko, Y. K. & Blau, W. J. (2004). High Performance Nanotube-Reinforced Plastics: Understanding the Mechanism of Strength Increase. *Advanced Functional Materials, 14*, 791–798.

[10] Kim, I. H. & Jeong, Y. G. (2010). Polylactide/Exfoliated Graphite Nanocomposites with Enhanced Thermal Stability, Mechanical Modulus, and Electrical Conductivity. *The Journal of Polymer Science: Polymer Physics, 48*, 850–858.

[11] Zhang, H. B., Zheng, W. G., Yan, Q., Yang, Y., Wang, J. W., Lu, Z. H., Ji, G. Y. & Yu, Z. Z. (2010). Electrically conductive polyethylene terephthalate/graphene nanocomposites prepared by melt compounding. *Polymer, 51*, 1191–1196.

[12] Chen, G., Wu, C., Weng, W., Wu, D. & Yan, W. (2003). Preparation of polystyrene/graphite nanosheet composites. *Polymer, 44*, 1781–1784.

[13] Ray, S. S. & Okamoto, M. (2003). Polymer/layered silicate nanocomposites: A review from preparation to processing. *Progress in Polymer Science, 28*, 1539–1641.

[14] Wang, Y. & Chen, W. C. (2013). Effect of clay modification on the dynamic mechanical and dielectric properties of PMMA nanocomposites via melt blending. *Polymer, 12*, 128–144.

[15] Ma, H., Tong, L., Xu, Z. & Fang, Z. (2007). Clay network in ABS-graft-MAH nanocomposites: Rheology and flammability. *Polymer Degradation and Stability, 92*, 1439–1445.

[16] Abraham, T. N., Ratna, D., Siengchin, S. & Karger-Kocsis, J. (2009). Structure and properties of polyethylene oxideorgano clay nanocomposite prepared via melt mixing. *Polymer Engineering & Science, 49*, 379–390.

[17] Kojima, Y., Usuki, A., Kawasumi, M., Okada, A., Fukushima, Y., Kurauchi, T. & Kamigiato, O. (1993). Mechanical properties of nylon 6-clay hybrid. *Journal of Materials Research, 8*, 1185–1189.

[18] Du, F., Fischer, J. E. & Winey, K. I. (2003). Coagulation method for preparing single-walled carbon nanotube/poly(methyl methacrylate) composites and their modulus, electrical conductivity, and thermal stability. *The Journal of Polymer Science: Polymer Physics, 41*, 3333–3338.

[19] Liu, L., Barber, A. H., Nuriel, S. & Wagner, H. D. (2005). Mechanical Properties of Functionalized Single-Walled Carbon-Nanotube/Poly(vinyl alcohol) Nanocomposites. *Advanced Functional Materials, 15*, 975–980.

[20] Shaffer, M. S. P. & Windle, A. H. (1999). Fabrication and characterization of CNT-PVA composites. *Advances in Materials, 11*, 937–941.

[21] Safadi, B., Andrews, R. & Grulke, E. A. (2002). Multiwalled carbon nanotube polymer composites: Synthesis and characterization of thin films. *Journal of Applied Polymer Science, 84*, 2660–2669.

[22] Chen, B., Evans, J. R. G., Greenwell, H. C., Boulet, P., Coveney, P. V., Bowdenf, A. A. & Whiting, A. (2008). A critical appraisal of polymer-clay nanocomposites. *Chemical Society Reviews, 37*, 568–594.

[23] Kuilla, T., Bhadra, S., Yao, D., Kim, N. H., Bose, S. & Lee, J. H. (2010). Recent advances in graphene polymer composites. *Progress in Polymer Science, 35*, 1350–1375.

[24] Aranda, P. & Ruiz-Hitzky, E. (1992). Poly(ethylene oxide)-silicate intercalation materials. *Chemistry of Materials, 4*, 1395–1403.

[25] Ruiz-Hitzky, E. & Aranda, P. (1990). Polymer-salt intercalation complexes in layer silicates. *Advances in Materials, 2*, 545–547.

[26] Shen, Z., Simon, G. P. & Cheng, Y. B. (2002). Comparison of solution intercalation and melt intercalation of polymer clay nanocomposites. *Polymer, 43*, 4251–4260.

[27] Liu, X. Q. & Chan-Park, M. B. (2009). Facile way to disperse single-walled carbon nanotubes using a noncovalent method and their reinforcing effect in poly(methyl methyacrylate) composites. *Journal of Applied Polymer Science, 114*. 3414–3419. Self Healing Graphene/ Polymer Composite – An Incisive Overview and its Recent Advancements 80.

[28] Kaminsky, W. & Funck, A. (2007). In situ polymerization of olefins with nanoparticles by metallocene-catalysis. *Macromolecular Symposia, 260*, 1–8.

[29] Kwon, S. M., Kim, H. S., Myung, S. J. & Jin, H. J. (2008). Poly(methyl methacrylate)/multiwalled carbon nanotubes microspheres fabricated via in-situ polymerization. *The Journal of Polymer Science: Polymer Physics, 46*, 182–189.

[30] Song, W. H., Ni, Q. P., Zheng, Z., Tian, L. Y. & Wang, X. L. (2009). The preparation of biodegradable polyurethane/carbon nanotube composite based on in situ crosslinking. *Polymers for Advanced Technologies, 20*, 327–331.

[31] Kwon, J. & Kim, H. (2005). Comparison of the properties of waterborne polyurethane/multiwalled carbon nanotubes and acid-treated multiwalled carbon nanotube composites prepared by in situ polymerization. *Journal of Polymer Science, Part A: Polymer Chemistry, 43*, 3973–3985.

[32] Mookhoek, S. D., Mayo, S. C., Hughes, A. E., Furman, S. A., Fischer, H. R. & Van der Zwaag, S. (2010). Applying SEM-based X-ray microtomography to observe self-healing in solvent encapsulated thermoplastic materials. *Advanced Engineering Materials, 12*, 228–234.

[33] Rule, J. D., Brown, E. N., Sottos, N. R., White, S. R. & Moore, J. S. (2005). Wax-protected catalyst microspheres for efficient self-healing materials. *Advances in Materials, 17*, 205–208.

[34] Das, S., Irin, F., Ma, L., Bhattacharia, S. K., Hedden, R. C. & Green, M. J. (2013). Rheology and morphology of pristine graphene/polyacrylamide gels. *ACS Applied Materials and Interfaces, 5*(17), 8633–8640.

[35] Thakur, S., Das, G., Raul, P. K. & Karak, N. (2013). Green one-step approach to prepare sulfur/ reduced graphene oxide nanohybrid for effective mercury ions removal. *The Journal of Physical Chemistry C, 117*(15), 7636–7642.

[36] Keller, M. W., White, S. R. & Sottos, N. R. (2007). A self-healing poly(dimethyl siloxane) elastomer. *Advanced Functional Materials, 17*, 2399–2404.

[37] Xiao, D. S., Yuan, Y. C., Rong, M. Z. & Zhang, M. Q. (2009). Hollow polymeric microcapsules: Preparation, characterization and application in holding boron trifluoride diethyl etherate. *Polymer, 50*, 560–568.

[38] Brown, E. N., Kessler, M. R., Sottos, N. R. & White, S. R. (2003). In situ poly(urea-formaldehyde) microencapsulation of dicyclopentadiene. *Journal of Microencapsulation, 20*, 719–730.

[39] Kim, J. T., Kim, B. K., Kim, E. Y., Kwon, S. H. & Jeong, H. M. (2013). Synthesis and properties of near IR induced self-healable polyurethane/graphene nanocomposites. *European Polymer Journal, 49*(12), 3889–3896.

[40] Suryanarayana, C., Rao, K. C. & Kumar, D. (2008). Preparation and characterization of microcapsules containing linseed oil and its use in self-healing coatings. *Progress in Organic Coatings, 63*, 72–78.

[41] Velev, O. D., Furusawa, K. & Nagayama, K. (1996). Assembly of latex particles by using emulsion droplets as templates 2. Ball-like and composite aggregates. *Langmuir, 12*.

[42] Yuan, Y. C., Rong, M. Z. & Zhang, M. Q. (2008). Preparation and characterization microencapsulated polythiol. *Polymer, 49*, 2531–2541.

[43] Yuan, Y. C., Ye, X. J., Rong, M. Z., Zhang, M. Q., Yang, G. C. & Zhao, J. Q. (2011). Self-healing epoxy composite with heat-resistant healant. *ACS Applied Materials Interfaces, 3,* 4487–4495.

[44] White, S. R., Sottos, N. R., Geubelle, P. H., Moore, J. S., Kessler, M. R., Sriram, S. R. . . . Viswanathan, S. (2001). Autonomic healing of polymer composites. *Nature, 409*(6822). 794–797.

[45] Yao, L., Rong, M. Z., Zhang, M. Q. & Yuan, Y. C. (2011). Self-healing of thermoplastics via reversible addition–fragmentation chain transfer polymerization. *The Journal of Materials Chemistry, 21,* 9060–9065.

[46] Yao, L., Yuan, Y. C., Rong, M. Z. & Zhang, M. Q. (2011). Self-healing linear polymers based on RAFT polymerization. *Polymer, 52,* 3137–3145.

[47] Yuan, L., Gu, A., Nutt, S., Wu, J., Lin, C., Chen, F. & Liang, G. (2013). Novel polyphenylene oxide microcapsules filled with epoxy resins. *Polymers for Advanced Technologies, 24,* 81–89.

[48] Suryanarayana, C., Rao, K. C. & Kumar, D. (2008). Preparation and characterization of microcapsules containing linseed oil and its use in self-healing coatings. *Progress in organic coatings, 63*(1), 72–78.

[49] Gragert, M., Schunack, M. & Binder, W. H. (2011). Azide/alkyne-"click"-reactions of encapsulated reagents: toward self-healing materials. *Macromolecular Rapid Communications, 32*(5).

[50] Patchan, M. W., Baird, L. M., Rhim, Y. R., LaBarre, E. D., Maisano, A. J., Deacon, R. M., Xia, Z. & Benkoski, J. J. (2012). Liquid-filled metal microcapsules. *ACS Appl Mater Interfaces, 4,* 2406–2412.

[51] Yuan, L., Liang, G. Z., Xie, J. Q. & He, S. B. (2007). Synthesis and characterization of microencapsulated dicyclopentadiene with melamine–formaldehyde resins. *Colloid and Polymer Science, 285,* 781–791.

[52] Ugelstad, J., El-Aasser, M. S. & Vanderhoff, J. W. (1973). Emulsion polymerization: Initiation of polymerization in monomer droplets. *Journal of Polymer Science, Polymer Letters Edition, 11,* 503–513.

[53] Bon, S. A. F., Mookhoek, S. D., Colver, P. J., Fischer, H. R. & Van der zwaag, S. (2007). Route to stable non-spherical emulsion droplets. *European Polymer Journal, 43.* 4839–4842. Self Healing Graphene/ Polymer Composite – An Incisive Overview and its Recent Advancements 83.

[54] Liu, X., Lee, J. K. & Kessler, M. R. (2009). Microencapsulation of self-healing agents with melamine-urea-formaldehyde by the Shirasu porous glass (SPG) emulsification technique. *Proceedings of SPIE, 7493*(749345).

[55] Yuan, L., Gu, A., Nutt, S., Wu, J., Lin, C., Chen, F. & Liang, G. (2013). Novel polyphenylene oxide microcapsules filled with epoxy resins. *Polymers for advanced technologies, 24*(1), 81–89.

[56] Xiao, D. S., Rong, M. Z. & Zhang, M. Q. (2007). A novel method for preparing epoxy-containing microcapsules via UV irradiation-induced interfacial copolymerization in emulsions. *Polymer, 48,* 4765–4776.

[57] Cui, J. & Del campo, A. (2012). Multivalent H-bonds for self-healing hydrogels. *Chemical Communication, 48,* 9302–9304.

[58] Maes, F., Montarnal, D., Cantournet, S., Tournilhac, F., Corté, L. & Leibler, L. (2012). Activation and deactivation of self-healing in supramolecular rubbers. *Software Matterials, 8,* 16817.

[59] Li, J., Hughes, A. D., Kalantar, T. H., Drake, I. J., Tucker, C. J. & Moore, J. S. (2014). Pickering-emulsion-templated encapsulation of a hydrophilic amine and its enhanced stability using poly (allyl amine). *ACS Macro Letters, 3*(10). 976–980.

[60] Whittell, G. R., Hager, M. D., Schubert, U. S. & Manners, I. (2011). Functional soft materials from metallopolymers and metallosupramolecular polymers. *Nature materials, 10,* 176–188.

[61] Liu, X., Le, J. K. & Kessler, M. R. (2011). Microencapsulation of self-healing agents with melamine-urea-formaldehyde by the Shirasu porous glass (SPG) emulsification technique. *Macromolecular Research, 19,* 1056–1061.

[62] Williams, K. A., Boydston, A. J. & Bielawski, C. W. (2007). Towards electrically conductive, selfhealing materials. *Journal of The Royal Society Interface, 4,* 359–362.

[63] Gun, A. K. & Mcdonald,. (2007). Exploring carbon flatland. *Physics Today, 60,* 35–41.

[64] Samandi, N. & Salezi, M. (2017). Self-healing and tough hydrogels with physically cross linked triple networked based on Agar/PVA/Graphene. *International Journal of Biological Macromolecules, 107*, 2291–2297.

[65] Silva, M. & Alves, N. (2017). M-graphene polymer composite for biomedical application. *Polymers for Advanced Technologies, 29*, 687–700.

[66] Hun, D. & Yan, L. (2016). Supramolecular hydrogel of chitosan in presence of graphene oxide nanosheet. *ACS Sustainable Chemistry & Engineering*, 302.

[67] Noack, M., Walter, A. & Benitez, A. (2017). Light fuelled spaciotemporal modulation of mechanical properties and rapid self-healing of graphene doped elastomer. *Advanced Functional Materials*, 27.

Chapter 4
Self-healing composition based on elastomers, thermoplastic elastomer (TPE), and thermoplastic vulcanizate (TPV)

4.1 Introduction

Elastomers are polymeric materials characterized by their ability to undergo large reversible deformations under stress [1]. This unique combination of high elasticity, flexibility, and resilience has led to their widespread use in industries ranging from automotive to aerospace, biomedical, consumer electronics, and construction [2]. However, like all polymeric materials, elastomers are inherently susceptible to mechanical damage, such as crack formation, fatigue-induced failure, and environmental degradation, which accumulate over time and compromise performance [1, 2].

The emergence of self-healing polymer technology offers a promising strategy to overcome these limitations [3]. Self-healing materials are designed to autonomously repair mechanical damage, restoring structural and functional integrity, without the need for external intervention or manual repair. By incorporating reversible chemical interactions or responsive physical mechanisms within the polymer matrix, these materials can detect and respond to damage stimuli, allowing the system to "heal" itself [3].

In the context of elastomeric systems, self-healing functionality is particularly attractive due to the intrinsic mobility of polymer chains, low glass transition temperatures (T_g), and amorphous network structures, all of which facilitate the dynamic rearrangement of bonding interactions necessary for healing [4, 5]. Furthermore, elastomers lend themselves to both intrinsic and extrinsic healing mechanisms, ranging from supramolecular assemblies and dynamic covalent bonds to encapsulated healing agents and thermally or mechanically activated healing systems [4].

Among elastomers, thermoplastic elastomers (TPEs) and thermoplastic vulcanizates (TPVs) have gained significant attention due to their hybrid structure: they combine the processability and recyclability of thermoplastics with the elastic recovery and resilience of cross-linked rubbers [6]. These materials are particularly suited for industrial processing (e.g., extrusion, injection molding, etc.) and offer a versatile platform for the development of next-generation self-healing elastomeric systems [6].

This chapter aims to provide a comprehensive overview of the current state of research and development in self-healing elastomer compositions, with particular focus on synthetic rubbers such as styrene–butadiene rubber (SBR), polyisoprene rubber (IR), isoprene isobutyl rubber (IIR), nitrile butadiene rubber (NBR), and chloroprene rubber (CR), along with the implementation of healing chemistries in TPEs and TPVs. Emphasis is placed on the underlying molecular mechanisms, network design strategies, and material performance metrics relevant to real-world applications.

https://doi.org/10.1515/9783111583716-004

4.2 Evolution of self-healing elastomers

The evolution of self-healing elastomers has closely mirrored the broader progression of polymer science and engineering, incorporating insights from multiple interdisciplinary domains such as supramolecular chemistry, dynamic covalent bonding, and bioinspired material design [7]. One of the earliest paradigms in this field involved the development of extrinsic self-healing systems, which laid the foundational groundwork for the conceptual realization of autonomous repair in synthetic polymer matrices. In these systems, healing functionality was not an inherent property of the polymer chains themselves, but rather was imparted through the inclusion of discrete healing agents, typically encapsulated in microcapsules, hollow fibers, or microvascular networks, distributed throughout the elastomeric material [7].

The operational mechanism of extrinsic self-healing elastomers relies on a stimulus-triggered release of these encapsulated agents. Upon the occurrence of mechanical damages such as a cut, crack, or puncture, the encapsulating shell would rupture under stress, liberating the internal healing agent (usually a low-viscosity monomer or reactive oligomer) into the damaged zone [8]. Once released, the healing agent would undergo in situ polymerization, often facilitated by an embedded catalyst or initiator system, also dispersed within the matrix. This localized chemical reaction leads to the formation of a polymeric patch or cross-linked network that partially or fully bridges the fracture site, thereby restoring the material's mechanical cohesion [8].

Despite the ingenuity of this approach and its proof-of-concept success in restoring partial functionality after damage, extrinsic systems suffer from a number of inherent limitations that restrict their long-term utility, especially in elastomeric applications that demand high fatigue resistance, flexibility, and dynamic strain tolerance [9]. One of the most critical drawbacks is their single-use nature. Once the encapsulated healing agent is depleted upon the first healing event, the material loses its capacity for further autonomous repair, making it unsuitable for applications that involve repeated mechanical stresses or cyclical loading.

Additionally, the physical and chemical integration between the healing agent and the host elastomer matrix is often suboptimal, leading to heterogeneous interfacial regions and incomplete recovery of mechanical properties [4]. The localized polymerization products may differ significantly in modulus, thermal expansion coefficient, or cross-link density compared to the original matrix, resulting in internal stress concentrations or new crack initiation sites. Another significant challenge lies in the restoration of complex morphologies. Extrinsic systems are generally incapable of reconstructing intricate or multiphase network architectures that are commonly found in engineered elastomeric materials, especially in the case of blends or composites [4, 5].

Moreover, extrinsic systems typically lack responsiveness to multidimensional stress fields, which is a key requirement in advanced elastomeric components used in aerospace, biomedical, or automotive applications, where materials are subjected to torsion, bending, shearing, and compression simultaneously [10]. As such, while extrinsic self-healing technologies provided an essential first step in demonstrating the feasibility of autonomous repair in synthetic polymers, they were ultimately found to be inadequate for the demanding and dynamic operational environments, character-istic of most elastomer applications [11]. This realization catalyzed a shift toward the development of intrinsic self-healing mechanisms, wherein the healing ability is em-bedded at the molecular level within the polymeric architecture itself – a topic ex-plored in greater detail in the subsequent sections of this chapter.

As the limitations of extrinsic approaches became evident, the research focus transitioned toward intrinsic self-healing systems [12]. These are based on reversible interactions that are integrated directly into the polymer backbone or cross-linking points. Elastomers are inherently well-suited for intrinsic self-healing due to their low glass transition temperatures and high segmental mobility, which enable polymer chains to move, interact, and reform bonds at damaged sites. Early intrinsic self-healing systems relied heavily on supramolecular interactions, such as hydrogen bonding, π–π stacking, host–guest chemistry, and metal–ligand coordination [2–5]. These physical cross-links provided dynamic, reversible interactions that could disso-ciate upon mechanical damage and subsequently reform, thereby enabling the mate-rial to recover its mechanical integrity. The entropic elasticity of elastomers further supported this mechanism by facilitating chain diffusion and network reorganization over time [13].

A major advancement in the field was the incorporation of dynamic covalent bonds into elastomeric matrices. Unlike supramolecular systems, which are primarily stabilized by physical interactions, dynamic covalent networks utilize chemical bonds that can break and reform under specific stimuli, such as heat, light, or pH. This ap-proach significantly enhanced the mechanical stability and long-term healing effi-ciency of self-healing elastomers [14]. Reversible chemistries such as disulfide bond exchange, imine and oxime formation, Diels–Alder and retro-Diels–Alder reactions, and transesterification reactions have been widely employed [15]. For example, disul-fide exchange offers redox-responsive or thermally activated healing pathways, while furan–maleimide Diels–Alder systems allow reversible cross-linking that can be tuned through thermal cycling. These chemistries enable elastomers to exhibit repeat-able healing behavior under controlled conditions, often achieving recovery of both tensile strength and elongation at break in successive damage-repair cycles [14, 15].

The latest phase in the evolution of self-healing elastomers has been driven by increasing industrial demand for reprocessable, recyclable, and sustainable materials [16]. This has TPEs and TPVs to the forefront of self-healing materials research. TPEs, typically consisting of block copolymers with alternating soft and hard segments, offer a microphase-separated morphology where the soft domains provide elasticity and the hard domains serve as physical cross-links [2–5]. The phase-separated nano-structure of TPEs permits segmental mobility at ambient or elevated temperatures, making them suitable for healing through physical rearrangement or thermally trig-gered reversible associations [17]. Similarly, TPVs – formed by dynamically vulcanizing rubber within a thermoplastic matrix – exhibit a fine dispersion of cross-linked elastomer particles within a continuous thermoplastic phase. This complex morphology allows for design strategies that promote localized healing at the rubber–matrix interfaces, particularly when dynamic cross-linkers or low-T_g modifiers are introduced into the elastomer phase [18].

Recent studies have demonstrated that processing methods such as reactive extrusion, in situ vulcanization, and melt blending can be tailored to control phase morphology and interface dynamics in TPE and TPV systems, enhancing their potential for intrinsic self-healing [19, 20]. Additionally, innovations in block copolymer architecture, filler compatibility, and interphase design are paving the way for the next generation of self-healing elastomeric materials, combining structural performance, healing efficiency, and sustainability [21].

Thus, the evolution of self-healing elastomers has progressed from extrinsic, single-use systems toward intrinsically dynamic, reprocessable materials with robust healing functionality. Continued advances in polymer chemistry, network architecture, and processing techniques will further expand the utility of self-healing elastomers in diverse and demanding applications.

4.3 Self-healing elastomer based on styrene–butadiene rubber

Styrene–butadiene rubber (SBR) is a synthetic copolymer widely used in elastomeric applications due to its excellent abrasion resistance, aging stability, and processability. Structurally, SBR comprises styrene (10–25%) and butadiene (75–90%) monomers, and its chemical framework allows various modifications to introduce functional groups or dynamic bonding sites [22]. One such modified derivative is carboxylated styrene–butadiene rubber (XSBR), which contains pendant carboxylic acid (–COOH) groups that facilitate the formation of reversible cross-linked networks [23]. These functionalities enable the design of intrinsically self-healing materials through ionic interactions or supramolecular assembly.

Xu et al. reported the development of self-healing XSBR by incorporating zinc oxide (ZnO), wherein Zn^{2+} ions form ionic cross-links by coordinating with the carboxylic acid groups of XSBR chains [24]. These metal–ligand interactions serve as reversible cross-linking points, and the material further forms secondary ionic networks through self-aggregation of Zn^{2+} ion pairs. Upon damage, the dynamic rearrangement of these ionic domains allows the restoration of network integrity, thereby imparting a self-healing capability and recyclability to the elastomer. Notably, XSBR containing 5 wt% ZnO exhibited a tensile strength of 6.7 MPa, which increased to 10.3 MPa after three recycling cycles, a performance significantly superior to most reported noncovalent supramolecular rubbers [24]. This mechanical robustness is attributed to the dual role of Zn^{2+}: initially forming primary ionic bonds and subsequently inducing the formation of ionic clusters that enhance both mechanical strength and healing performance.

Further advancement in ionic supramolecular networks was demonstrated by Xu et al. through the preparation of XSBR/nano-chitosan (NCS) composites. In this system, the $-NH_2$ groups of NCS engage in salt-forming reactions with the $-COOH$ groups of XSBR, resulting in the formation of carboxylate-ammonium ion pairs ($[COO^-][NH_3^+]$). These ion pairs further self-assemble into weak ionic clusters, forming a dynamic supramolecular hybrid network that imparts reversible cross-linking behavior and enables self-healing (Fig. 4.1). The incorporation of 20 wt% NCS significantly enhanced the mechanical strength, with the tensile strength of the XSBR/NCS composite doubling that of pristine XSBR. Additionally, the healing efficiency reached 92% at room temperature after 24 h. However, at higher NCS loadings (30–40 wt%), the healing efficiency decreased due to the formation of dense filler–filler networks, which restricted polymer chain mobility and hindered the interfacial reorganization necessary for effective healing. Thus, NCS serves as a multifunctional component, contributing both to cross-linking and mechanical reinforcement, provided its loading remains within the optimal range (Fig. 4.1).

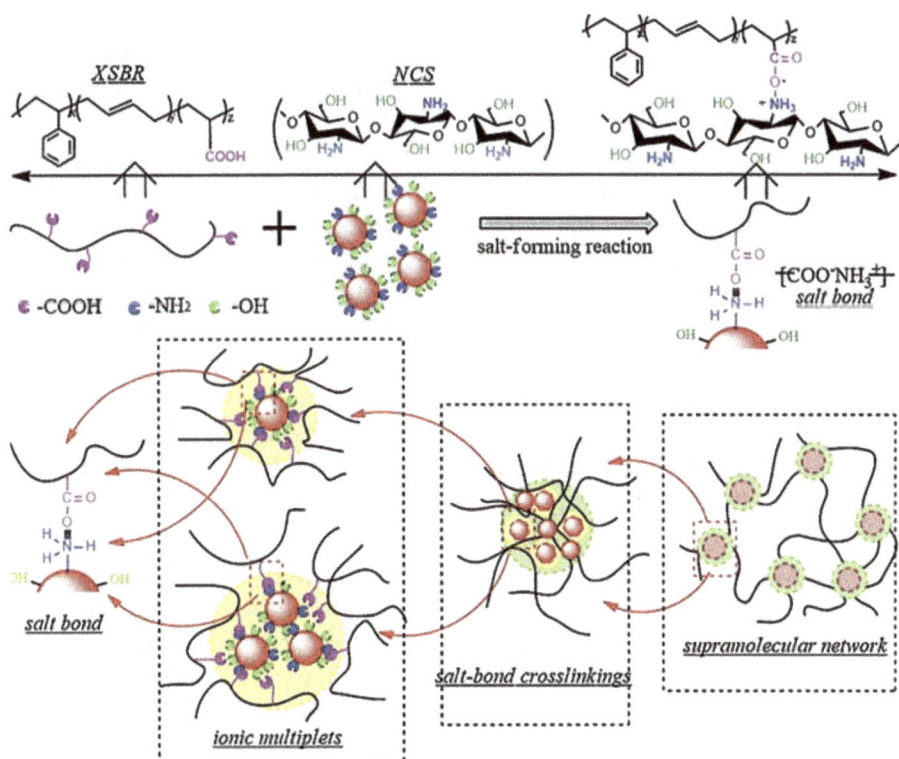

Fig. 4.1: Schematic illustration of salt-forming reaction between XSBR and NCS and evolution of supramolecular hybrid network in XSBR/NCS composites, reprinted with permission [25], copyright reserved Elsevier 2019.

In addition to ionic and supramolecular strategies, covalent reversible bonding mechanisms have also been explored to develop self-healing SBR systems. Kuang et al. reported the fabrication of SBR/CNT nanocomposites utilizing Diels–Alder (DA) and retro-Diels–Alder (rDA) chemistry. In this system, SBR was modified with furfuryl groups (SBR-FS), and multi-walled carbon nanotubes were functionalized with maleimide-terminated furfuryl groups (MWCNT-FA) [46]. These components underwent DA reactions with bismaleimide (BM) to form a covalently cross-linked, yet thermally reversible, elastomeric network. The incorporation of 5 wt% MWCNT-FA led to a 200–300% increase in Young's modulus, attributed to both an increase in cross-link density and the reinforcing effect of the nanotubes. The healing performance was highly dependent on the stoichiometric ratio of furan to maleimide groups. At a furan/maleimide ratio of 3:1, the healing efficiency increased from 37% (without MWCNT-FA) to 90% with 5 wt% MWCNT-FA, upon heating at 100 °C for 5 h. However, when the ratio was adjusted to 1:1, the healing efficiency decreased with increased MWCNT-FA loading due to reduced chain mobility from excessive cross-linking. Comparatively, composites with COOH-functionalized MWCNTs (MWCNT-COOH) did not exhibit comparable healing properties, highlighting the importance of specific functionalization that can participate in reversible DA chemistry. These findings suggest that MWCNT-FA serves a dual function: forming covalent reversible cross-links and acting as a structural reinforcement agent.

Another promising approach to achieve self-healing behavior in SBR involves dynamic covalent chemistry based on polysulfide metathesis reactions. In this context, low molecular weight polysulfide rubber was blended with SBR to introduce dynamic polysulfide linkages capable of undergoing reversible exchange reactions. The SBR/PSR blend was compounded with conventional rubber ingredients and vulcanized using metal chloride catalysts, which facilitated polysulfide bond exchange within the cross-linked network. The resulting elastomer displayed a healing efficiency of up to 87% along with improved tensile strength, indicating the successful formation of a dynamic and reprocessable network [27]. However, the introduction of silica filler into the SBR/PSR composite reduced the healing efficiency, likely due to the constrained segmental mobility imposed by the filler network. This observation shows a common trade-off in filled self-healing systems, where mechanical enhancement through rigid fillers can adversely affect the dynamic reorganization required for healing [27].

Overall, the self-healing capabilities of SBR-based elastomers can be effectively realized through a variety of mechanisms, including reversible ionic coordination, dynamic supramolecular networks, thermally reversible covalent chemistry, and dynamic covalent polysulfide bonding. Each approach presents unique advantages and limitations in terms of healing efficiency, mechanical strength, reprocessability, and temperature requirements. The choice of dynamic chemistry and filler type must be judiciously balanced to tailor the performance of the self-healing elastomer for specific applications such as flexible electronics, automotive components, or sustainable rubber materials.

4.4 Self-healing elastomer based on polyisoprene rubber

Polyisoprene rubber, encompassing both natural rubber (NR) and synthetic isoprene rubber (IR), is primarily composed of cis-1,4-polyisoprene units [28]. The high cis-1,4 content imparts superior elasticity, tensile strength, and fatigue resistance, particularly in NR, whose biosynthetic pathway yields a highly regular stereochemistry, not easily replicated in synthetic variants [28]. This structural regularity and unsaturation within the hydrocarbon backbone provide excellent mechanical properties but also make the polymer susceptible to oxidative degradation and thermal instability. Furthermore, the lack of intrinsic dynamic bonding sites limits its capacity for autonomous damage recovery, making self-healing behavior largely inaccessible in its pristine state [29].

To overcome these limitations, significant efforts have been directed toward the chemical functionalization of polyisoprene chains with dynamic, reversible crosslinkable moieties, which can respond to external stimuli such as heat, pressure, or moisture [30]. Among the most promising strategies is the introduction of ionic crosslinks, which offer reversible, noncovalent interactions capable of re-forming after mechanical damage [31].

For instance, Yang et al. developed a novel self-healing NR-based ionomer by reacting brominated NR with histidine, a bio-inspired ligand containing an imidazole group. The imidazole–carboxylate ionic pairing between the functional groups facilitates the formation of a physically cross-linked supramolecular network. This network exhibits rapid self-healing behavior, achieving ~10 s repair cycles and healing efficiencies as high as 93.98% at room temperature. The healing process is driven by the reversible nature of ionic associations, which promote molecular mobility and reorganization at the damaged interface [32].

Despite these advantages, elastomers that rely solely on ionic cross-linking often suffer from notable limitations in their mechanical performance [33]. Specifically, they tend to exhibit inadequate tensile strength, reduced toughness, and poor dimensional stability when subjected to prolonged mechanical stress, deformation, or environmental exposure [33]. These shortcomings arise because ionic bonds, while reversible and beneficial for self-healing, are inherently weaker and more susceptible to disruption under extreme conditions compared to their covalent counterparts [34].

To overcome these issues, researchers have developed hybrid elastomeric systems that incorporate both ionic and covalent cross-linking within the same polymer network. In these dual cross-linked architectures, the ionic interactions serve a crucial role in facilitating dynamic and reversible bonding, which enables the material to self-heal after mechanical damage or microstructural failure [34]. Meanwhile, the covalent cross-links contribute permanent, strong connections that maintain the overall structural integrity and provide the mechanical robustness necessary for load-bearing applications [33, 34].

This synergistic combination leverages the strengths of both bonding mechanisms: the adaptability and healing capacity of ionic cross-links, and the durability and resilience of covalent bonds. As a result, the resulting elastomers are capable of maintaining their shape, mechanical properties, and functionality across multiple damage-healing cycles without significant degradation in performance [35]. This advancement significantly expands the practical applicability of polyisoprene-based self-healing elastomers, making them suitable for more demanding environments, including those encountered in advanced engineering systems, wearable electronics, soft robotics, and biomedical devices.

Xu et al. employed a controlled peroxide-induced vulcanization approach to integrate zinc dimethacrylate (ZDMA) into NR, thereby forming a supramolecular network, primarily stabilized by ionic cross-links [36]. By retarding conventional covalent cross-link formation during vulcanization, the material maintained dynamic reversibility. When cut and rejoined, these materials recovered their mechanical properties nearly to the original levels at room temperature, attributed to the re-formation of ZDMA-based ionic clusters. In a variation of this system, in situ formation of ZDMA was achieved by compounding NR with zinc oxide (ZnO) and methacrylic acid (MAA) (Fig. 4.2). This generated an ionic hybrid network during peroxide curing, and the residual nano-ZnO served both as a reinforcing filler and a cross-linking facilitator. The system exhibited enhanced mechanical properties; however, healing efficiency declined with increased temperature due to degradation of labile ionic domains. The ZnO/MAA ratio was found to be critical – higher ratios improved mechanical and healing efficiencies but retarded the healing kinetics due to stiffer, more tightly bound networks.

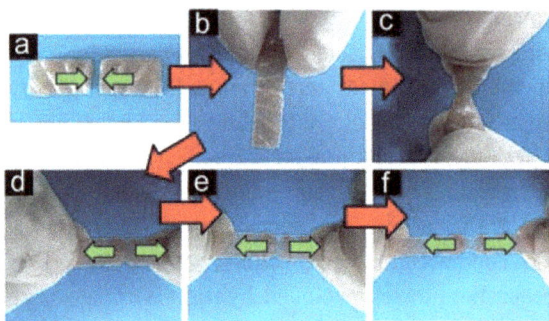

Fig. 4.2: Photographs of 60 s-cured NR/ZDMA compound illustrating self-healing behavior at room temperature: (a) cut strip-shape sample; (b) healed for 5 min at room temperature to be integrity; (c) healed sample under strong twisting; (d, e, and f) healed sample under manual stretching, reprinted with permission [36], copyright reserve American Chemical Society 2016.

Further enhancement in both mechanical performance and healing efficiency was demonstrated by Xu et al. through a dual ionic network approach. Initially, lithiated

NR was reacted with maleic anhydride to introduce lithium carboxylate groups, forming the first dynamic network. This was followed by a secondary network formation via polymerization of ZDMA under peroxide curing. The resultant material exhibited superior toughness and self-healing performance when compared to singly cross-linked analogs, emphasizing the synergistic benefit of combining different ionic motifs within the elastomer matrix [38].

Beyond simple ionic interactions, dynamic covalent chemistries – particularly disulfide metathesis – have emerged as a powerful tool for imparting reversible bonding characteristics to epoxidized natural rubber (ENR) systems. Disulfide bonds are of particular interest due to their thermo-reversible nature, allowing for network rearrangement and material reprocessability under elevated temperatures [39]. Among the notable approaches, Imbernon et al. demonstrated the incorporation of aliphatic disulfide bonds into ENR through the use of dithiodibutyric acid (DTDB) as a bifunctional cross-linker. Rheological characterization confirmed that DTDB is a highly efficient cross-linker, exhibiting cross-linking kinetics and final mechanical moduli comparable to those achieved with conventional non-dynamic diacid cross-linkers such as dodecanedioic acid (DA) [39].

Upon heating from ambient temperature to 180 °C, both DTDB- and DA-cross-linked systems displayed a sharp increase in the elastic modulus (G′), indicating effective cross-link formation. Swelling tests in toluene revealed similar cross-link densities between the two systems, supporting the assertion that both cross-linkers form structurally comparable networks via epoxy ring opening and formation of β-hydroxyester linkages [39]. However, the unique feature of the DTDB-cross-linked network lies in its dynamic covalent nature. Reductive swelling experiments using tri-n-butylphosphine (TBP) demonstrated that disulfide bonds remain intact post-curing and can be selectively cleaved, leading to significant network disassembly and dissolution. This confirms the presence and accessibility of disulfide linkages in the cured material, a property absent in the DA-cured control [39].

Importantly, these disulfide bonds facilitate significant stress relaxation at elevated temperatures (180 °C), a behavior not observed in the DA-based network [40]. The capacity for stress relaxation under constant strain highlights the potential for network rearrangement via reversible bond exchange, further distinguishing DTDB as a dynamic cross-linker. While the presence of side reactions at high temperatures – such as radical-induced coupling between thiyl radicals and ENR double bonds – can lead to irreversible cross-links (e.g., monosulfide bonds), the dominant dynamic behavior is retained over multiple cycles, indicating the formation of a predominantly reversible disulfide-based network with a minor permanent component [41].

Further creep/recovery tests and cyclic loading experiments confirmed the thermo-adaptive nature of the DTDB-cured ENR [39]. At room temperature (30 °C), the material exhibited elastic behavior and full shape recovery, consistent with conventional rubber-like properties. However, at elevated temperatures (150–180 °C), enhanced creep deformation was observed, attributed to thermally activated disulfide exchange. The time-dependent slowing of creep rate at 180 °C suggests progressive suppression of reversibility due to side reactions, emphasizing a trade-off between thermal activation and long-term network stability [39].

Extending this concept, Cheng et al. employed aromatic disulfide cross-linkers such as DTSA and DTDA in combination with sulfur to fabricate dual cross-linked ENR systems [42]. These materials leveraged both dynamic disulfide exchange and thermally labile hydrogen bonding to achieve healing efficiencies approaching 98% at 120 °C. However, an excess of sulfur led to over-cross-linking, which hindered chain mobility and significantly reduced healing performance [42].

Hydrogen bonding and reversible metal–ligand coordination are important classes of noncovalent dynamic interactions that contribute to self-healing behavior in isoprene rubber systems. These interactions do not form permanent chemical bonds but can break and reform under suitable conditions, allowing materials to repair damage.

In one study, Liu et al. introduced a dual dynamic cross-linking strategy that combined hydrogen bonding with metal–ligand coordination to improve the self-healing ability and mechanical recovery of polyisoprene. The system incorporated Zn^{2+} ions complexed with triazole groups and multiple hydrogen-bonding sites [43]. This created a network in which both types of reversible interactions worked together to enhance elasticity and enable reformation after damage. When tested, the material recovered 74% of its original tensile strength after being heated to 80 °C, showing that the combination of these dynamic interactions can lead to a mechanically robust and reprocessable elastomer [43].

In another approach, researchers modified cellulose nanocrystals (CNCs) by attaching histidine groups, which can coordinate with Zn^{2+} ions. These modified CNCs were then mixed with epoxidized natural rubber (ENR) that had also been functionalized with $ZnCl_2$. The histidine–Zn^{2+} coordination formed a dynamic network within the rubber, while the CNCs contributed additional mechanical reinforcement [44]. This system exhibited exceptionally high healing efficiency (over 100%) in the first few cycles, likely due to the combination of reversible coordination and physical reinforcement from the nanocrystals. However, the tensile strength of the material was low (around 1 MPa), indicating that while the material could repair itself efficiently, its structural durability was limited. This illustrates a trade-off, often seen in supramolecular systems: increasing reversibility and healing often comes at the cost of mechanical strength [44].

Polyisoprene has also been functionalized for Diels–Alder chemistry and transesterification reactions [45]. DA-based networks typically require elevated temperatures (~130 °C) for healing and prolonged post-treatment at ~40 °C to regenerate the DA adducts for mechanical stabilization. Conversely, transesterification networks based on polyisoprene require even higher healing temperatures (>160 °C), limiting their practical application. Although these dynamic covalent systems offer high healing efficiencies and superior thermal stability, their stringent thermal requirements reduce their applicability in ambient or low-temperature conditions [46].

Among the many chemical modification strategies applied to natural rubber, epoxidized natural rubber stands out due to its dual reactive nature. ENR contains both epoxy groups and carbon–carbon double bonds along its polymer backbone, as we revealed [49]. This structural feature introduces two distinct types of chemical reactivity: the epoxy groups can undergo nucleophilic ring-opening reactions or form reversible physical interactions such as hydrogen bonding, while the double bonds can participate in free radical cross-linking or other addition reactions [50]. The presence of these two functional moieties allows ENR to support a wide range of dynamic bonding mechanisms, making it suitable for self-healing applications that require tunable cross-link density and reversible network formation.

Experimental studies have shown that even unmodified ENR exhibits some degree of self-healing capability when exposed to elevated temperatures [51]. This limited healing behavior has been attributed primarily to chain mobility and interdiffusion across damaged regions. As temperature increases, polymer chains at the fracture interface gain sufficient mobility to partially re-entangle or diffuse into each other. Additionally, the epoxy groups may form weak, reversible interactions across the interface, further contributing to healing. However, these effects are generally modest in the absence of additional network design elements [52, 53].

To improve the healing performance, Rahman et al. investigated the effects of blending ENR with ionomers, specifically ethylene–methacrylic acid copolymers neutralized with either sodium (EMNa) or zinc ions (EMZn) [54]. They observed that ENR blended with EMNa exhibited nearly complete healing of mechanical properties after damage, whereas the EMZn blend was less effective. This difference in performance is attributed to the behavior of the metal cations. Sodium ions tend to form weaker, more reversible ionic interactions with carboxylate groups in the ionomer, facilitating rapid reorganization of the network after damage. In contrast, zinc ions form stronger, more static coordination bonds that may impede the dynamic rearrangement necessary for efficient healing [54].

Thus, the self-healing performance of ENR-based materials is governed by a complex interplay of chain mobility, dynamic bond reformation, and interfacial compatibility. Epoxidation level, choice of ionic or coordination agents, curing strategy, and the nature and functionalization of nanofillers, each play critical roles in determining the kinetics and extent of healing. Achieving a balance between rapid network reformation and mechanical reinforcement remains a central challenge in designing high-performance self-healing elastomers based on ENR.

4.5 Self-healing elastomer based on isoprene isobutyl rubber (IIR)

IIR is a synthetic elastomer derived from the cationic copolymerization of isobutylene with a minor amount of isoprene, typically in the range of 1–3 mol% [55]. The polymerization is conducted under cryogenic conditions (−100 to −80 °C) using Lewis acid catalysts such as aluminum chloride or boron trifluoride. The resulting copolymer exhibits a largely saturated backbone, composed primarily of isobutylene repeat units, which imparts low gas permeability, high chemical resistance, and thermal stability [56]. The small amount of isoprene introduces isolated double bonds, primarily as methyl-substituted internal olefins that are chemically accessible for subsequent functionalization. These unsaturations are strategically important because they allow for halogenation or cross-linking, while remaining sparse enough to preserve the base polymer's oxidative and thermal robustness [55].

Bromobutyl rubber (BIIR) is produced by selective electrophilic bromination of these allylic sites within the isoprene units. The process involves reaction of IIR with elemental bromine or brominating agents under controlled conditions to prevent excessive degradation or overfunctionalization [57]. Bromine atoms are introduced at the allylic positions, yielding BIIR with approximately 1.5–2 wt% bromine content. The resulting allylic bromides are highly reactive due to stabilization of the transition state by adjacent π systems, making them efficient substrates for nucleophilic substitution reactions. These reactive sites are key enablers for post-synthetic modifications that allow tailored incorporation of functional moieties [58].

One particularly effective route for chemical modification involves the use of N-substituted imidazoles as nucleophiles. When reacted with BIIR, these imidazoles displace the allylic bromide, resulting in quaternized imidazolium salts tethered to the polymer backbone. Bromide serves as the counterion (Fig. 4.3) [59]. The installation of these ionic imidazolium groups introduces electrostatically interactive sites along the elastomer chains. These sites spontaneously organize into ionic aggregates via Coulombic interactions. The clusters formed by these interactions act as physical, noncovalent cross links, which are both thermodynamically and kinetically reversible under moderate thermal conditions [60].

The dynamic nature of these ionic clusters is central to the self-healing behavior observed in these modified materials. Upon mechanical failure or fracture, the ionic clusters at the damage interface are disrupted [61]. Heating the material increases the segmental mobility of the chains and promotes dissociation of the ionic aggregates, allowing polymer chains from opposing fracture surfaces to interdiffuse [62]. Upon cooling, the ionic groups reassociate, reconstituting the physical network structure and thereby restoring the mechanical properties. This process is autonomous in the sense that it does not require external intervention beyond thermal stimulus and interfacial contact, and it leverages inherent ionic interactions as a dynamic cross-linking mechanism.

The efficiency and robustness of this self-healing process are closely related to the structure of the alkyl group on the imidazole ring used in the modification [63]. Studies have shown that short-chain substituents, such as methyl groups, yield strong ionic aggregates due to high local charge density and minimal steric interference [64]. These tightly associated ionic domains restrict the polymer chain mobility and significantly inhibit the interdiffusion necessary for healing, although they confer high mechanical strength. Conversely, longer alkyl chains such as butyl, hexyl, or nonyl introduce steric hindrance and increase free volume, which weakens the electrostatic association between adjacent ionic groups. This loosening of the cluster network enhances segmental motion and diffusion, thereby improving the healing response under thermal activation [65].

Among these, 1-hexylimidazole emerges as the most effective modifier for achieving a favorable balance between mechanical performance and healing efficiency. BIIR modified with this compound exhibits a healing efficiency of 74%, when assessed by recovery of tensile strength, and 98% based on elongation at break. The absolute mechanical values after healing were reported as 10.7 MPa for tensile strength and 1,040% for elongation at break, indicating that the network structure re-establishes to a significant degree post healing. The observed balance suggests that the cluster strength and chain mobility must be carefully tuned to optimize both material toughness and reversibility.

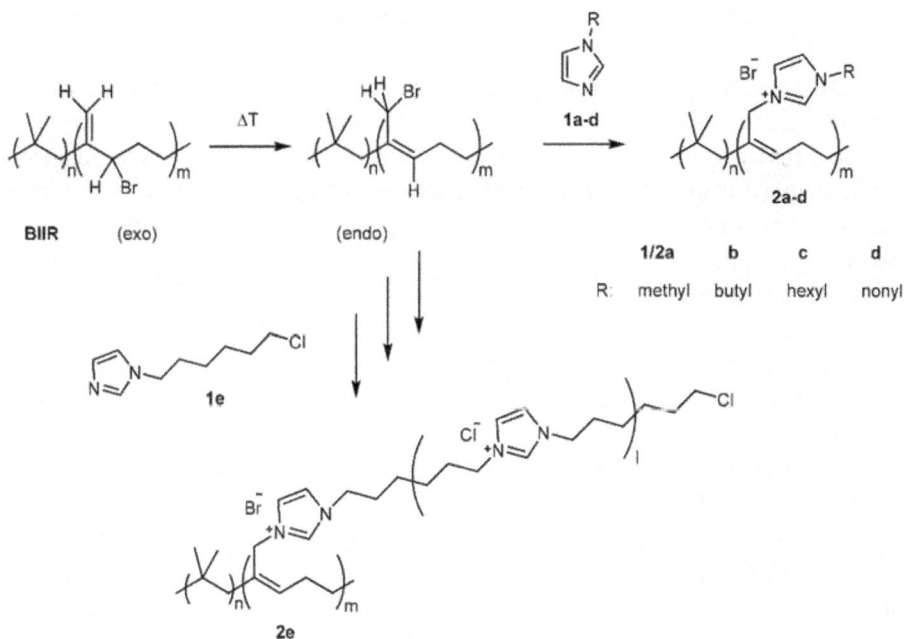

Fig. 4.3: Ionic modification of bromobutyl rubber by conversion with alkylimidazoles, reprinted with permission from [66], copyright reserved American Chemical Society 2017.

Recently, the self-healing behavior of BIIR modified with 1-butylimidazole (IM) has been investigated with the aim of developing advanced elastomeric composites for high-performance applications. The modification introduces imidazolium ionic moieties into the BIIR matrix via nucleophilic substitution at the allylic bromide positions, facilitating the formation of reversible ionic clusters that function as dynamic physical cross-links [67].

To further improve the multifunctional performance of the composite, 1-butyl-3-methylimidazolium bromide, a thermally stable and highly conductive ionic liquid, was incorporated as a plasticizing and ionic transport medium. Its inclusion enhances the mobility of ionic species and facilitates dynamic interfacial interactions within the polymer network, contributing to both self-healing efficiency and electrical conductivity [67].

The elastomeric matrix was formulated as a binary blend of BIIR and natural rubber (NR) in two distinct phr ratios: 70:30 and 80:20 (BIIR:NR). Natural rubber, with its inherent elasticity and strain-induced crystallization behavior, complements the impermeability and chemical resistance of BIIR. This synergistic blending optimizes the viscoelastic and mechanical performance of the resulting materials. To further reinforce the composites, a hybrid filler system comprising carbon nanotubes and carbon black (CB) was incorporated [67]. These conductive fillers not only improve the mechanical integrity and modulus of the composite but also establish percolated conductive pathways, enabling electro-responsive healing and recyclability.

The focus of the work is to create a class of ionic–elastomeric materials that exhibit rapid and efficient self-healing under electrical stimulation, with superior curing kinetics, enhanced mechanical and dynamic-mechanical properties, and stable electrical conductivity. The puncture stress recovery was used as a key parameter to assess the mechanical self-healing efficiency after damage. Electrical recyclability tests were performed to examine the composite's ability to restore conductivity following damage–repair cycles, thereby demonstrating functional recovery of both structural and conductive networks [67].

In the same context, self-healing anticorrosive coatings designed for operation in chemically aggressive environments have emerged as a critical area of research, particularly in fields such as marine engineering, aerospace, and the oil and gas industries, where materials are routinely exposed to extreme pH and high salinity. Thus, BIIR presents a versatile platform due to its inherent chemical resistance, low gas permeability, and modifiable allylic bromide functionality. In the present work, a novel ionically cross-linked system was developed by incorporating 4-tert-butylpyridine (TBP) into the BIIR matrix. The nitrogen lone pair on the TBP readily engages in nucleophilic substitution with the brominated sites on BIIR, forming pyridinium-based ionic cross-links. These ionic interactions are electrostatically driven and thermally reversible, enabling dynamic network reconstruction upon damage [68]. The design leverages the low glass transition temperature of BIIR, which facilitates chain mobility at ambient or mildly elevated temperatures, allowing ionic clusters to dissociate and re-form. This dynamic ionic bonding, in combination with the hydrophobic and sterically hindered structure of TBP, provides the coating with both flexibility and chemical resilience. As a result, the TBP–BIIR coating demonstrates robust self-healing capabilities and superior anticorrosive properties across a wide range of aggressive environments.

Electrochemical impedance spectroscopy (EIS) measurements reveal that the TBP–BIIR coating achieves high protection efficiency values under severe corrosive conditions: 94.37% in strong acidic media (HNO$_3$, pH = 2), 94.95% in alkaline environments (NaOH, pH = 12), and 95.49% in saline conditions (3.5 wt% NaCl aqueous solution). These results indicate that the ionic bonding between TBP and the BIIR matrix remains intact and effective across various ionic strengths and pH extremes, a property rarely achieved with conventional polymer coatings [68].

Notably, the coating's ability to self-heal after mechanical damage was also quantified. Upon mechanical failure and subsequent exposure to the environment, the damaged coatings spontaneously reformed their ionic cross-linked structure. The repaired films maintained a PE exceeding 89% across all tested conditions, with maximum post-healing PE values approximating 95%. This highlights the exceptional reversibility of the ionic interactions and the rapid reconstitution of the barrier function.

As we have learnt, IIR, when chemically modified via nucleophilic substitution with imidazole derivatives, undergo a transformation in their network structure, effectively behaving as ionically cross-linked elastomers. This ionic functionalization introduces dynamic supramolecular interactions in place of traditional covalent cross-links [69].

Moreover, the resulting ionically cross-linked elastomeric system displays significant alterations in its physical and mechanical performance profile [61]. The introduction of reversible ionic interactions not only enhances thermal and mechanical stability but also imparts a self-healing capability under mild external stimuli, such as modest heating or prolonged rest at room temperature [70]. This self-repair behavior is a direct consequence of the dynamic nature of the ionic clusters, which can dissociate and re-form, enabling the material to recover from microstructural damage. Scanning electron microscopy has been employed to directly observe the healing of fractured surfaces, revealing that the surface topology post healing closely resembles the original undamaged morphology (Fig 4.4).

In-depth mechanical characterization including stress–strain analysis, dynamic mechanical analysis, and fatigue crack growth measurements has demonstrated that the imidazole-modified BIIR exhibits superior tear resistance and enhanced fatigue life, relative to its unmodified or conventionally cross-linked counterparts [61]. These properties show the effectiveness of ionic association in maintaining network integrity under cyclic or prolonged mechanical stress. Furthermore, the elimination of curatives and high-temperature vulcanization steps streamlines rubber processing, reducing energy consumption and the number of synthetic stages required. This simplification of manufacturing not only lowers production costs but also mitigates the generation of toxic byproducts typically associated with sulfur vulcanization chemistry. As such, this approach contributes to the development of more environmentally benign elastomeric materials [71].

Importantly, the versatility of the ionic cross-linking mechanism permits the extension of this methodology to other halogen-functionalized elastomers [72], broadening the scope of potential applications. The inherent tunability of the imidazole sub-

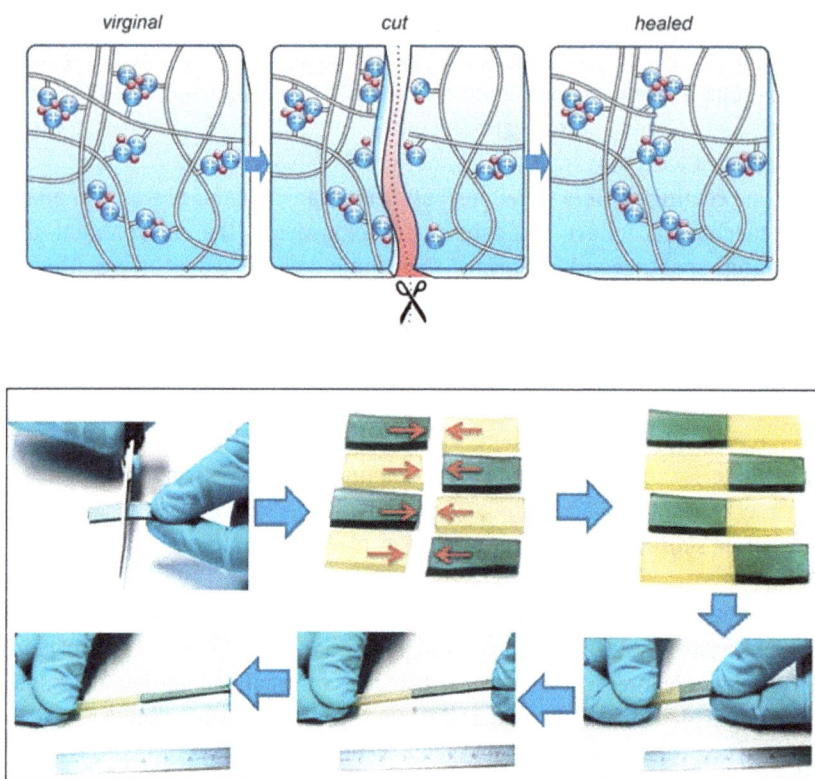

Fig. 4.4: Ionic association and self-healing. (Top) Schematic representation of self-healing by re-formation of ionic associates in an ionic rubber network. (Bottom) A sample of imidazolium-modified bromobutyl rubber was cut into two pieces. After reassembling and 18 h of healing time, the sample was stretched to breakdown. For better visualization, one part of the cut sample was colored, reprinted with permission from [61], copyright reserved American Chemical Society 2015.

stituents and their influence on the strength and reversibility of ionic clusters offers material scientists a powerful tool for tailoring mechanical, thermal, and healing properties to meet specific performance criteria [73].

Additionally, the mechanical robustness of these elastomers can be further enhanced by compounding with traditional reinforcing fillers such as carbon black or silica [74, 75]. The compatibility of the modified BIIR with such fillers has been demonstrated, allowing the formulation of composites that meet the demanding requirements of structural applications, while retaining self-healing capabilities [75].

Nevertheless, the broad applicability of this ionic self-healing strategy to other halide-containing polymers suggests a generalizable methodology for developing next-generation elastomeric and protective materials. The ease of chemical transformation, absence of toxic curatives, and potential for recycling and reprocessing provide compelling reasons to consider these ionomeric elastomers as replacements for traditional rubbers in advanced engineering contexts.

4.6 Self-healing elastomer based on nitrile butadiene rubber (NBR)

Nitrile butadiene rubber (NBR) is a copolymer that contains both unsaturated carbon–carbon double bonds and polar nitrile functionalities. The nitrile group, in particular, plays a crucial role in enabling coordination chemistry with various metal ions [76]. These interactions are not limited to simple complexation but extend to network formation, where the metal ions act as cross-linking agents via coordination with the nitrile groups [77]. This mode of cross-linking introduces a form of reversibility, not typically observed in covalently cross-linked systems. When the coordination bond is dynamic and reversible, the resulting network becomes capable of thermal reprocessing. This characteristic makes coordination-cross-linked NBR potentially recyclable, which is of growing interest in sustainable materials development [78].

The use of functionalized NBRs further expands the scope of coordination cross-linking. For instance, the incorporation of catechol functionalities via dopamine grafting introduces additional metal-binding sites. Cheng et al. demonstrated this by preparing a recyclable NBR network through the epoxidation of NBR, followed by dopamine functionalization [79]. Subsequent complexation with Fe^{3+} ions formed a network in which the catechol–metal coordination bonds acted as reversible cross-links. This system was not only recyclable but also showed promise for self-healing, suggesting that the dynamic nature of the coordination bond can support recovery of mechanical integrity after damage [79].

In a related study, Zhang et al. investigated the use of cobalt neocaprate (cobalt(II) decanoate) as a coordination cross-linker for nitrile butadiene rubber, revealing a significant enhancement in self-healing behavior when compared to conventional sulfur-cured systems [80]. The coordination-cross-linked NBR demonstrated a healing efficiency of 72% after 10 min at 190 °C, a substantial improvement over the 26% efficiency observed in its sulfur-cross-linked counterpart under the same conditions. This contrast underscores the distinct advantages of dynamic metal–ligand coordination bonds over traditional covalent cross-links, which are largely irreversible under typical processing conditions. The reversible nature of coordination bonds allows the polymer network to temporarily dissociate and reorganize when exposed to elevated temperatures, facilitating the reformation of interfacial bonds at damaged regions [80].

The effectiveness of cobalt neocaprate in this context lies in its ability to form relatively strong yet thermally labile coordination bonds with the nitrile groups of NBR [81]. These bonds act as reversible cross-links that not only allow for self-repair upon heating but also maintain a coherent network structure during regular use. Intriguingly, the mechanical properties of the cobalt-cross-linked NBR, particularly tensile

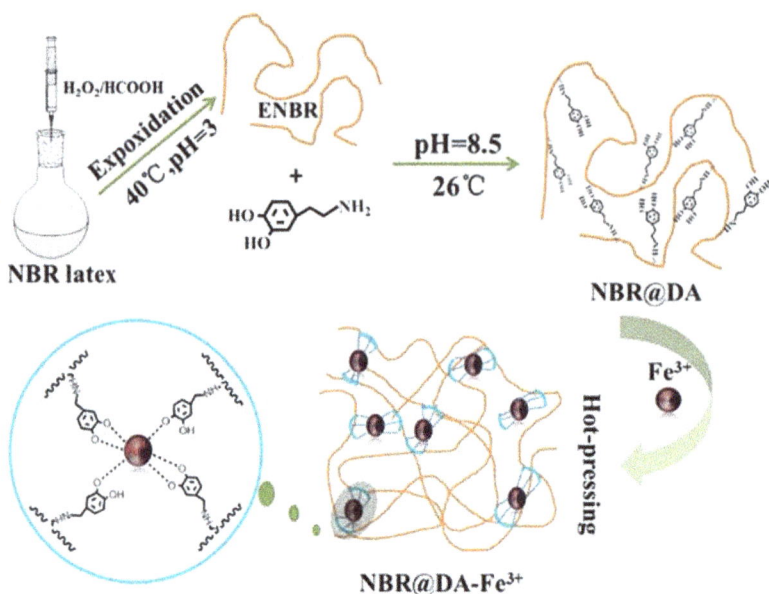

Fig. 4.5: Illustration of the synthesis process of the modified NBR samples, reprinted with permission from [79], copyright reserved American Chemical Society 2019.

strength and modulus, were not compromised by the dynamic nature of the cross-linking but were in fact improved relative to sulfur-cured samples, suggesting that the cobalt–nitrile coordination imparts a dual benefit: it reinforces the material through localized network structuring, while preserving the capacity for network mobility and rearrangement under thermal stimuli.

This outcome, however, also exposes an inherent tension between network stability and dynamic reconfigurability. Stronger coordination bonds, such as those involving cobalt(II), can enhance mechanical robustness, but they may simultaneously restrict the molecular mobility required for efficient self-healing. This is due to a decrease in the entropy of the system and a suppression of the chain dynamics necessary for effective diffusion and re-association across damaged interfaces. Thus, while the material gains in structural integrity, its healing performance may plateau or even decline if the metal–ligand interactions become too stable [80, 81].

The observed performance of the cobalt-cross-linked NBR therefore exemplifies a nuanced optimization challenge in the design of dynamic polymer networks: the metal–ligand coordination strength must be sufficiently high to sustain mechanical demands, yet not so high as to suppress the reversible dynamics that underpin healing. Future work in this area may involve tuning ligand denticity, incorporating mixed-metal systems, or modulating ligand field strength to fine-tune this balance. Furthermore, time–temperature–healing correlations need to be carefully mapped to understand the kinetics of network dissociation and re-formation, which remain critical for real-world application scenarios such as tire rubbers, seals, or impact-resistant coatings, where both strength and durability are demanded in conjunction with repairability.

The identity of the coordinating metal plays a decisive role in determining the balance between mechanical strength and healing capacity. Das et al. investigated this in the context of DAP-functionalized carboxylated NBR (DAP-XNBR), coordinated with Co^{2+}, Ni^{2+}, and Zn^{2+} salts. While Zn^{2+}-coordinated samples showed superior self-healing efficiency, Co^{2+} and Ni^{2+} coordinated networks displayed higher tensile strengths [80]. The authors attributed this to the different dynamic natures of the metal–ligand interactions. Zn^{2+} forms more labile bonds, enabling reversible association and dissociation conducive to healing, whereas Co^{2+} and Ni^{2+} form stronger, less dynamic bonds that restrict segmental motion.

Other approaches to enhancing the self-healing capability of nitrile butadiene rubber involve blending it with auxiliary polymers or additives that introduce or amplify dynamic interactions within the matrix. One such strategy was reported by Schüssele et al., who investigated the incorporation of hyperbranched polyethyleneimine (PEI) into NBR. PEI is a polyamine-based polymer containing primary, secondary, and tertiary amine groups, along with urea functionalities when modified accordingly [83]. These chemical groups are capable of engaging in hydrogen bonding, electrostatic interactions, and possibly reversible covalent chemistry under appropriate conditions. The three-dimensional, highly branched architecture of hyperbranched PEI introduces a high density of functional end groups, which significantly increases the number of potential interaction sites within the NBR network [84].

The study found that blending NBR with 12 phr of high-molecular-weight PEI (2000 g/mol) markedly improved healing efficiency, reaching values up to 44%, compared to a baseline of 4% for unmodified NBR. In contrast, blends with low-molecular-weight PEI (800 g/mol) achieved a lower healing efficiency, suggesting that molecular weight plays a decisive role in establishing a percolated network of dynamic interactions [83]. This observation can be attributed to the greater chain length and entanglement capability of higher molecular weight PEI, which not only increases the density of interaction sites but also enhances chain mobility and stress redistribution near the damage site. Moreover, the urea and amine functionalities may also act as hydrogen bond donors and acceptors, contributing to reversible physical cross-links that can re-form during healing. Thus, the introduction of hyperbranched PEI into NBR does not merely add an inert filler but functions as a dynamic healing promoter by enhancing both the physical and chemical responsiveness of the system.

In a separate but thematically related study, Barrios et al. explored the incorporation of poly(acrylic acid)-grafted ground tire rubber (PAA-g-GTR) into carboxylated nitrile rubber (XNBR), cross-linked with zinc oxide (ZnO). In the XNBR/ZnO system, Zn^{2+} ions coordinate with the carboxyl groups present on the polymer chains to form ionic cross-links. These ionic interactions are inherently more labile than covalent bonds, allowing for partial dissociation and re-formation when the system is subjected to external stimuli such as heat or mechanical stress. However, the baseline healing efficiency of this system remains relatively modest, with the unmodified XNBR/ZnO compound achieving only 15% healing under standard test conditions [85].

The introduction of PAA-g-GTR dramatically alters this behavior. By grafting poly (acrylic acid) chains onto recycled ground tire rubber, additional acid groups are introduced into the system. These groups can interact with Zn^{2+} ions to form new ionic clusters or reinforce existing ones, thereby increasing both the density and spatial distribution of reversible cross-linking sites. As a result, the healing efficiency of the modified system increases substantially, reaching up to 70%. This dramatic improvement illustrates the critical importance of not merely increasing cross-link density per se, but optimizing the cross-linking network to support dynamic reorganization [85]. The mobility and reversibility of ionic clusters formed between Zn^{2+} ions and acid groups allow damaged interfaces to re-associate effectively during thermal treatment. The presence of PAA-g-GTR may also serve a dual function: on the one hand, it provides chemical functionality that promotes healing, and on the other, it contributes physical reinforcement from the rubber filler phase, helping to maintain mechanical integrity during and after the healing process [86].

Together, these studies highlight a central principle in the design of self-healing NBR systems: introducing dynamic, multivalent interactions, whether through physically blended polymers like PEI or chemically functionalized fillers like PAA-g-GTR, which can significantly improve the material's ability to recover after damage. However, these improvements are highly dependent on the molecular architecture, functional group compatibility, and mobility of the introduced species, which must be carefully tuned to balance mechanical strength and healing efficiency. Future work may explore synergistic effects from combining multiple such additives, or investigate the healing kinetics in relation to the microstructure of the blend, to further optimize performance for practical applications.

A more sophisticated approach involves dual dynamic chemistries, as reported by Sarkar et al., who engineered XNBR systems, functionalized with furan groups, capable of undergoing Diels–Alder reactions (Fig. 4.6). These were further cross-linked with BMDS, a bifunctional cross-linker containing both maleimide and disulfide units. This dual-functional network leveraged both thermos-reversible DA bonds and photo-responsive disulfide exchange to enable healing under different stimuli. Healing efficiencies ranged from 60% to 91%, depending on the temperature, UV exposure, and cross-linker concentration. The authors noted that increasing the cross-link density improved tensile strength at the expense of healing ability, due to reduced chain mobility. While their study dissected the contributions of individual stimuli (thermal vs. UV), it stopped short of analyzing their synergistic effect, a missed opportunity that could have further clarified the interplay between cross-link dynamics and network reconfiguration [87].

Thus, the self-healing capacity of NBR is closely tied to the nature of its cross-linking strategy. Reversible coordination bonding, dynamic covalent interactions, and

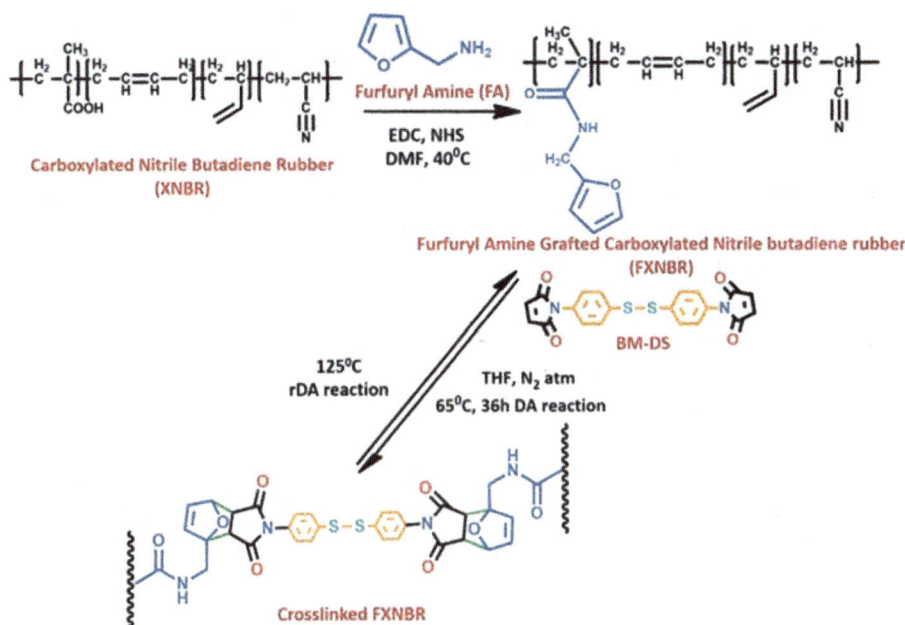

Fig. 4.6: Modification of XNBR with furfuryl amine (FA) and Diels–Alder cross-link of modified XNBR (FXNBR), reprinted with permission from [87], copyright reserved Wiley 2021.

multifunctional additives, each offer different mechanisms for damage recovery. However, a critical balance must be struck between enhancing mechanical performance and preserving sufficient molecular mobility for healing. These findings point to the need for a nuanced approach to network design, where the dynamic behavior of the cross-linking chemistry is as important as the static properties of the material.

4.7 Self-healing elastomers based on chloroprene rubber (CR)

Chloroprene rubber (CR) is traditionally vulcanized using sulfur-based systems, resulting in a highly cross-linked network with strong covalent bonds that resist deformation, flow, or repair once set [88]. This makes CR durable in service, but entirely unsuitable for reshaping or healing after damage. To overcome this limitation, recent research has focused on introducing dynamic covalent chemistry into the network, particularly through disulfide exchange reactions. These reactions allow certain types of sulfur cross-links to break and re-form under controlled conditions, thereby enabling limited reprocessability and self-healing without complete structural failure [89].

In one study, Xiang et al. introduced copper(II) methacrylate (MA-Cu) into sulfur-cross-linked CR to act as a catalyst for disulfide metathesis. The catalyst activates the disulfide bonds when the rubber is heated to 120 °C [90]. At this temperature, bond exchange reactions occur both in the bulk material and at the damaged surfaces, allowing the material to heal or be reshaped. Below this temperature, typically below 100 °C, the catalyst remains inactive, and the material retains its mechanical stability. The system works by targeting disulfide and polysulfide bonds within the network. However, monosulfide bonds, common byproducts of conventional vulcanization, are inert under these conditions and cannot undergo exchange. As these bonds accumulate, they reduce the dynamic behavior of the network and limit the material's ability to heal or be reused. Thus, the effectiveness of this approach depends heavily on controlling the type of sulfur bonds formed during curing [90].

A different strategy involves building the reversibility directly into the network from the outset, rather than relying on a catalyst added post-curing. In a recent study, a liquid polysulfide polymer, Thiokol LP-3, was used as a cross-linker and chemically bonded to the CR backbone via a thiol-ene reaction. This reaction was thermally initiated, forming covalent bonds between thiol groups of the cross-linker and unsaturated sites on the polymer chain. Unlike traditional sulfur vulcanization, this method introduces disulfide bonds as a structural component of the network itself, which can later undergo exchange without requiring an external catalyst [91].

Three formulations were prepared with increasing concentrations of Thiokol LP-3 – 2, 4, and 6 parts per hundred rubber (phr). Mechanical testing showed that the sample with 2 phr of cross-linker had the highest tensile strength (22.4 MPa) and the greatest elongation at break (16.2%). Increasing the cross-linker concentration to 4 phr led to reduced strength (11.7 MPa) and elongation (10.7%), while 6 phr resulted in even lower values (5.6 MPa and 7.3% strain). These results suggest that increasing the amount of cross-linker increases the overall cross-link density, which restricts chain mobility and reduces the material's ability to deform without breaking. This higher stiffness can be beneficial for certain applications but comes at the cost of flexibility and toughness [91].

Healing efficiency, in contrast, increased with cross-linker concentration. When damaged samples were held at 80 °C for 24 h, healing efficiencies of 4.5%, 4.4%, and 13.4% were measured for the 2, 4, and 6 phr samples, respectively. The improved healing performance with higher cross linker content likely reflects the increased number of dynamic bonds capable of exchange. However, this improvement in healing comes at the expense of mechanical performance. Moreover, the absolute healing efficiency remains relatively low, indicating that while some network recovery occurs, the extent of bond re-formation is still limited possibly due to slow exchange kinetics or reduced chain mobility in the more densely cross-linked samples [91].

Recyclability was also evaluated. The formulation with 4 phr of Thiokol LP-3 showed the best balance, achieving a recycling efficiency of 11.2%. This suggests that at intermediate cross-linker content, there is a sufficient number of dynamic bonds to allow partial network rearrangement while preserving enough mechanical strength to support structural integrity after reprocessing [91].

Together, these studies demonstrate two contrasting approaches to enabling self-healing in CR. The first relies on adding a catalyst that activates only certain types of sulfur cross-links under heat, and its success is limited by the composition of the original vulcanized network. The second uses a designed cross-linker to embed dynamic bonds into the network structure from the start, giving more control over the healing behavior but requiring changes to the formulation and curing process. Both methods show promise but also highlight the need to balance reversibility with mechanical performance. High healing efficiency often requires more mobile, loosely cross-linked networks, while good mechanical properties rely on strong, stable cross-links. The challenge lies in tuning these competing factors to meet the requirements of the intended application.

4.8 Some special references regarding the implementation of TPE-based self-healing polymers

TPEs uniquely combine the elasticity of rubber with the processability of thermoplastics due to their microphase-separated morphology [92, 93]. This morphology is characterized by hard, physical cross-linking domains dispersed in a continuous soft elastomeric matrix, enabling TPEs to sustain elastic deformation while allowing reversible network rearrangement under appropriate conditions. Exploiting this reversible network architecture is central to implementing self-healing functionality in TPE systems [94].

The fundamental self-healing mechanism in conventional TPEs is intrinsically linked to their unique block copolymer structure, which drives microphase separation into chemically distinct hard and soft domains [95]. This phase separation arises due to the incompatibility between the chemically different polymer blocks, resulting in a morphology where discrete hard segments aggregate into physically cross-linked domains embedded within a continuous elastomeric matrix [96].

The hard domains are typically composed of glassy or semicrystalline polymers such as polystyrene (PS) or poly(methyl methacrylate) (PMMA). These hard blocks possess a relatively high glass transition temperature (T_g) or melting temperature (T_m), which enables them to act as physical cross-linking points. These cross-links provide the material with its mechanical strength, dimensional stability, and shape memory, essentially functioning as reversible anchors that maintain the integrity of the polymer network under stress [97].

In contrast, the soft domains consist of elastomeric polymer blocks, for example polybutadiene (PB), polyisoprene (PI), or ethylene–propylene rubber (EPR), characterized by low glass transition temperatures well below room temperature. These soft blocks impart elasticity to the TPE, allowing for significant reversible deformation. They provide the polymer chains with the necessary mobility to undergo conformational changes and flow at elevated temperatures or under mechanical force [95–97].

When mechanical damage occurs, such as the initiation of microcracks, surface scratches, or clean cuts, the integrity of the physical network is compromised as the continuous elastomer matrix and the discrete hard domains become disrupted. This disruption results in a loss of mechanical strength and the formation of fracture surfaces [98].

Self-healing in these materials is primarily activated by heating the TPE above the softening or glass transition temperature of the hard domains. Upon reaching these temperatures, the hard domains undergo a phase transition where their glassy or semicrystalline structure softens or partially dissociates [97]. This transition weakens the physical cross-links, increasing the polymer chain mobility substantially at the damaged interface. Increased mobility enables the polymer chains on the opposing fracture surfaces to diffuse across the interface and re-entangle with one another [97]. As the material cools back below the T_g or T_m of the hard domains, these domains re-solidify and the physical cross-links reform, effectively re-establishing the network structure and restoring the mechanical integrity of the material. The re-formation of these hard domains "locks" the chains into place, sealing the damage and achieving healing.

The success and efficiency of this healing process are influenced by multiple factors. The extent of polymer chain diffusion across the damaged interface is directly affected by the temperature, relative to the softening transitions of the hard and soft blocks [99]. Higher temperatures facilitate greater chain mobility but must be controlled to avoid material degradation or loss of mechanical properties. Additionally, the molecular weight of the polymer chains influences chain diffusion; higher molecular weights generally reduce mobility, requiring longer times or higher temperatures for effective healing [100].

Polymer architecture, such as linear, branched, or block copolymer design, also governs the ease with which chains can diffuse and re-entangle. The size, shape, and connectivity of the phase-separated domains critically affect the healing kinetics. Smaller, well-dispersed hard domains with sufficient interfacial area promote more efficient healing by facilitating chain interpenetration. In contrast, large or poorly connected domains may inhibit chain diffusion and reduce healing efficiency.

Despite these advantages, the requirement of elevated temperatures, often exceeding 100 °C to achieve sufficient chain mobility for healing, presents a significant limitation. This temperature constraint restricts the applicability of purely thermally activated self-healing TPEs in many real-world scenarios, particularly where in situ or ambient-temperature healing is desired. As a result, ongoing research focuses on incorporating dynamic covalent or supramolecular bonds within TPE networks to enable healing at lower temperatures and improve the practicality of self-healing thermoplastic elastomers.

To overcome the temperature limitations and improve healing efficiency, modern implementations introduce dynamic covalent or supramolecular bonds into the TPE network. Unlike purely physical cross-links, these reversible chemical bonds can break and re-form under mild conditions, allowing healing to occur at lower temperatures and shorter timescales [101].

Dynamic covalent chemistries frequently used in TPE systems include disulfide metathesis, Diels–Alder cycloaddition, imine exchange, and boronic ester formation [102]. These chemistries enable network rearrangement through bond exchange reactions, which can be thermally or photo-triggered. For example, disulfide bonds can undergo metathesis reactions upon heating, enabling chain rearrangement while maintaining network connectivity. Similarly, Diels–Alder adducts can reversibly dissociate and reform upon temperature changes, providing thermal control over healing [101].

In addition, supramolecular interactions such as hydrogen bonding, metal–ligand coordination, and host–guest chemistry have been utilized to impart dynamic cross-linking in TPEs [103]. These interactions are generally weaker than covalent bonds but highly reversible and can respond to external stimuli such as temperature, pH, or mechanical stress. For instance, multiple hydrogen bonding sites, grafted along polymer chains, increase the density of transient cross-links, enabling rapid network recovery upon damage. Metal–ligand coordination bonds offer tunable bond strength, depending on the choice of metal ion and ligand, which directly influences healing kinetics and mechanical properties [104].

A critical consideration in the implementation of self-healing TPEs is balancing the network's mechanical robustness with its healing capacity. Increasing the cross-link density or the strength of physical and chemical interactions enhances mechanical strength, modulus, and thermal stability but concurrently reduces chain mobility, which is essential for effective healing [101].

Conversely, reducing the cross-link density or weakening the reversible interactions improves chain diffusion and healing efficiency but can compromise tensile strength, elasticity, and durability [105]. Therefore, material design requires systematic optimization of block lengths, molecular weights, cross-linker concentration, and dynamic bond density to strike an optimal balance for specific applications [106].

Furthermore, the kinetics of bond exchange and the lifetime of dynamic bonds affect both the speed and completeness of healing. Faster bond dynamics facilitate rapid healing but may reduce the material's mechanical integrity under load, whereas slower kinetics improve mechanical stability but extend the healing time. Advanced polymer synthesis and processing techniques enable tuning these parameters by controlling polymer architecture and incorporating functional groups at precise locations within the chain.

Elucidating the self-healing behavior and mechanisms in TPEs requires comprehensive characterization techniques that assess both mechanical performance and microstructural changes. Dynamic mechanical analysis provides insights into the viscoelastic properties and transitions of phase-separated domains, essential for understanding chain mobility and network stability during healing [107].

Atomic force microscopy and transmission electron microscopy enable visualization of the phase morphology and its evolution after damage and healing. Spectroscopic methods such as Fourier-transform infrared spectroscopy and X-ray photoelectron spectroscopy can detect changes in chemical bonding associated with dynamic covalent or supramolecular interactions [101].

Mechanical testing, including tensile, fracture toughness, and fatigue tests, quantifies the restoration of properties post-healing. Repeated damage and healing cycles are performed to evaluate the durability and longevity of the self-healing function. Rheological studies provide data on bond-exchange kinetics and network dynamics relevant to healing efficiency.

Self-healing TPEs present a compelling solution in applications where mechanical flexibility, long-term durability, and damage resilience are essential. These materials combine the processability of thermoplastics with the elastic behavior of cross-linked rubbers, making them well-suited for use in sectors that demand both mechanical performance and structural adaptability [101].

In flexible electronics, self-healing TPEs can address one of the major limitations of wearable devices – mechanical failure due to repetitive strain, bending, or minor physical damage [108, 109]. By enabling autonomous or thermally induced repair of microcracks and interfacial delamination, TPEs help maintain device performance and longevity without requiring user intervention or full replacement. The same principles apply in stretchable conductors, soft robotics, and electronic skins, where material failure can compromise both function and safety [109].

In the automotive and aerospace industries, self-healing TPEs offer a route toward reducing lifetime maintenance costs and material waste. TPEs already see wide use in interior components, seals, and vibration-dampening elements [110]. Incorporating self-healing functionality in these applications means that superficial scratches, stress-induced microcracks, and environmental degradation can be partially or fully reversed without requiring part replacement. The ability of certain self-healing TPEs to recover structural and functional properties under moderate heating makes them compatible with localized repair protocols already used in these industries [110,111].

Sealants, gaskets, and protective coatings also benefit from TPE-based self-healing formulations. These materials often fail due to the accumulation of small mechanical flaws or chemical exposure. Introducing reversible bonding motifs (e.g., supramolecular hydrogen bonding, metal–ligand coordination, or dynamic covalent linkages such as Diels–Alder adducts) into the TPE matrix enables the material to maintain seal integrity and surface continuity over extended periods, especially under fluctuating mechanical or environmental stress [92].

A specific subclass of TPEs, thermoplastic vulcanizates (TPVs), deserves particular attention in the context of self-healing. TPVs are dynamic composites made by dispersing cross-linked rubber particles (commonly EPDM or NBR) within a thermoplastic matrix (often polypropylene), typically through dynamic vulcanization during melt mixing. This architecture creates a co-continuous or phase-separated morphology where the elastomeric phase provides elasticity, and the thermoplastic matrix grants thermal reprocessability [112]. While TPVs are already used in automotive weather seals, hoses, and vibration-control parts, their implementation in self-healing systems introduces new challenges and opportunities.

The key difficulty with TPVs is that the rubber phase is typically covalently cross-linked and thus inert to reorganization under mild conditions. To enable self-healing, the vulcanization chemistry must be re-engineered to incorporate dynamic covalent bonds (e.g., disulfides, Diels–Alder adducts, oxime esters) or reversible supramolecular interactions. In practice, this means replacing traditional sulfur-based cross-linking agents with dynamic cross-linkers or grafting reversible functional groups onto the rubber phase. The thermoplastic matrix, in turn, can act as a facilitator of chain mobility, upon heating, supporting chain diffusion across fractured interfaces [113,114].

For example, introducing disulfide-containing cross-linkers into the elastomer phase of a TPV enables healing through thermally induced bond reshuffling [92]. This could allow partial recovery of mechanical integrity without compromising the thermoplastic nature of the matrix. Alternatively, embedding hydrogen-bonding motifs in both phases allows healing at lower temperatures, albeit with typically reduced mechanical strength. The fine balance between mobility (required for healing) and stiffness (required for performance) remains a central design consideration [115].

Despite these promising developments, major challenges remain before self-healing TPEs and TPVs can be broadly deployed. First, achieving efficient healing at ambient or physiological temperatures without the need for prolonged heating, pressure, or solvent exposure remains a barrier for many applications. Second, repeated healing cycles often lead to diminished mechanical performance due to incomplete network re-formation, fatigue damage accumulation, or irreversible chemical degradation. Third, scalability of synthesis routes – particularly for TPEs incorporating dynamic chemistries – remains limited by cost, synthetic complexity, or the need for specialized catalysts.

Ongoing research efforts are focusing on advanced polymer architectures that combine multiple dynamic chemistries, hierarchical self-assembly, and controlled phase morphologies. For instance, block copolymers with dual-responsive motifs (e.g., temperature- and pH-sensitive domains) can enable tunable healing behavior [116–118]. Additive strategies using small-molecule healing promoters, nanoparticles, or compatibilizers are also being explored to enhance the healing response without compromising mechanical integrity or thermoplasticity.

In conclusion, the self-healing capabilities of TPEs and TPVs derive from the interplay between their phase-separated morphology and the reversible interactions embedded within their structure. Through careful molecular design – particularly in the dynamic vulcanization chemistry of TPVs and thorough characterization of thermal, mechanical, and healing behavior – these systems can evolve into robust, reprocessable materials with significantly extended functional lifetimes. Their implementation aligns closely with sustainable materials engineering goals, where damage tolerance and recyclability are key performance metrics.

4.9 Conclusion and outlook

The field of self-healing elastomeric materials has witnessed an extensive evolution over the past two decades, evolving from basic, single-use extrinsic systems into highly sophisticated, intrinsically dynamic polymer networks [92]. In the early stages, self-healing was primarily achieved through the inclusion of microencapsulated healing agents or vascular delivery networks systems that could only heal once and lacked long-term utility in demanding, cyclic environments [30]. These designs laid the conceptual foundation for autonomous repair but fell short in durability and repeatability. The current generation of self-healing elastomers overcomes these limitations through molecular-level engineering that integrates reversible interactions directly into the polymer backbone or cross-link junctions, enabling repeatable, stimuli-responsive healing under ambient or controlled conditions [40, 41].

This scientific progression reflects a broader societal and industrial imperative to develop materials that are not only durable and high-performing but also sustainable and recyclable [62]. With increasing emphasis on circular economy principles and reduced environmental impact, materials capable of extending service life, minimizing maintenance, and reducing waste are in high demand. Self-healing elastomers address these needs by autonomously repairing damage, thus reducing the frequency of replacements and associated resource consumption. Moreover, the incorporation of re-processable chemistries aligns well with green manufacturing strategies and energy-efficient processing routes [92].

At the molecular level, the key enabler of this evolution has been the strategic implementation of dynamic reversible interactions within the elastomer network. These interactions, ranging from physically reversible hydrogen bonding and π–π stacking to chemically reversible covalent bonds such as disulfide exchange, imine formation, and Diels–Alder adducts, allow the polymer matrix to respond adaptively to mechanical damage. The choice of healing mechanism often dictates the performance envelope of the material, with trade-offs between healing speed, mechanical strength, and operational temperature range [101].

Different classes of elastomers have embraced these strategies in diverse ways. Styrene–butadiene rubber has leveraged both ionic coordination and Diels–Alder chemistry to form reversible networks with impressive tensile recovery and recyclability. Polyisoprene rubber has benefited from bio-inspired ionic cross-links and disulfide metathesis, providing healing efficiencies near 90% under mild conditions. Isoprene isobutyl rubber, particularly in its brominated form, has been functionalized with imidazolium-based ionic groups to create thermally reversible networks, yielding promising results for anticorrosive and electrical applications. Similarly, nitrile butadiene rubber has demonstrated high healing potential through dynamic metal–ligand coordination and supramolecular blending with hyperbranched polymers or poly(acrylic acid)-modified fillers. Chloroprene rubber, traditionally rigid due to sulfur vulcanization, has shown self-healing capabilities through engineered disulfide-containing cross-linkers and catalytic activation.

While each of these systems offers unique advantages, a central challenge remains in balancing network rigidity (which confers mechanical strength) with segmental mobility (which promotes healing). Reversible supramolecular networks typically allow rapid healing at low energy input but often compromise strength, whereas covalent dynamic networks, though stronger, may require higher temperatures or specific triggers to activate healing.

Amid these developments, thermoplastic elastomers and thermoplastic vulcanizates have emerged as especially compelling platforms for self-healing innovation. Their microphase-separated morphology – consisting of hard thermoplastic domains and soft elastomeric matrices – naturally supports both structural resilience and dynamic reconfiguration. Through careful molecular design, including the introduction of dynamic covalent bonds or supramolecular motifs into either phase, TPEs and TPVs can be tailored to exhibit self-healing under thermal, mechanical, or electrical stimuli. Moreover, their compatibility with melt processing methods such as extrusion, injection molding, and 3D printing enhances their industrial scalability and commercial readiness.

Nevertheless, several technological challenges must be addressed for self-healing elastomers to transition from laboratory demonstrations to widespread practical applications. Chief among these is the need to achieve efficient healing at ambient or physiological temperatures without sacrificing long-term mechanical stability. Healing systems must also retain their performance over multiple damage–repair cycles without cumulative degradation [119]. Additionally, the synthesis of dynamic networks must be streamlined to allow cost-effective, reproducible scaling for mass production. The development of predictive design tools and real-time monitoring techniques to assess healing behavior in situ will further accelerate this transition.

Looking ahead, the future of self-healing elastomers lies in the development of hybrid architectures that synergistically combine multiple dynamic interactions, nanostructured fillers, and hierarchical phase morphologies. These next-generation materials may feature dual-responsive networks, compartmentalized healing zones, or bioinspired hierarchical designs, capable of adapting to complex, real-world loading conditions. By integrating these innovations, it is conceivable that elastomeric materials will not only match but surpass traditional rubbers in durability, functionality, and environmental compatibility.

Nevertheless, self-healing elastomeric materials, particularly those engineered from TPE and TPV frameworks, represent a transformative approach to materials design. Their ability to autonomously restore structural and functional integrity not only reduces life cycle costs but also advances sustainability goals across a wide spectrum of applications. From automotive components and aerospace seals to flexible electronics and biomedical implants, these smart materials herald a new era of resilient, reconfigurable, and eco-conscious polymer technology.

References

[1]   Shanks, R. A. & Kong, I. (2013). General purpose elastomers: structure, chemistry, physics and performance. *Advances in Elastomers I: Blends and Interpenetrating Networks*, 11–45.
[2]   Chen, Q., Liang, S. & Thouas, G. A. (2013). Elastomeric biomaterials for tissue engineering. *Progress in Polymer Science*, *38*(3–4), 584–671.
[3]   Idumah, C. I. (2021). Recent advancements in self-healing polymers, polymer blends, and nanocomposites. *Polymers and Polymer Composites*, *29*(4), 246–258.
[4]   Roy, N., Bruchmann, B. & Lehn, J. M. (2015). Dynamers: dynamic polymers as self-healing materials. *Chemical Society Reviews*, *44*(11), 3786–3807.
[5]   Li, B., Cao, P. F., Saito, T. & Sokolov, A. P. (2022). Intrinsically self-healing polymers: from mechanistic insight to current challenges. *Chemical Reviews*, *123*(2), 701–735.
[6]   Naskar, K. & Babu, R. R. (2015). Thermoplastic elastomers (TPEs) and thermoplastic vulcanizates (TPVs). In: Kobayashi, S. & Müllen, K., (eds.), Encyclopedia of Polymeric Nanomaterials (pp. 2517–2522).
[7]   Harun-Ur-Rashid, M., Jahan, I., Islam, M. J., Kumer, A., Huda, M. N., Imran, A. B. . . . Alsalhi, S. A. (2024). Global advances and smart innovations in supramolecular polymers. *Journal of Molecular Structure*, *1304*, 137665.
[8]   Pezzin, S. H. (2023). Mechanism of extrinsic and intrinsic self-healing in polymer systems. In: Multifunctional Epoxy Resins: Self-Healing, Thermally and Electrically Conductive Resins (pp. 107–138). Singapore: Springer Nature Singapore.
[9]   Wang, X., Sedaghati, R., Rakheja, S. & Shangguan, W. (2025). Rubber fatigue revisited: a state-of-the -art review expanding on prior works by tee, mars and fatemi. *Polymers*, *17*(7), 918.
[10] Sahu, B. B., Moharana, S. & Behera, P. K. (2024). Elastomeric-based composite materials for engineering applications. In: Polymer Composites: Fundamentals and Applications (pp. 329–355). Singapore: Springer Nature Singapore.
[11] Almutairi, M. D., Aria, A. I., Thakur, V. K. & Khan, M. A. (2020). Self-healing mechanisms for 3D-printed polymeric structures: From lab to reality. *Polymers*, *12*(7), 1534.
[12] Choi, K., Noh, A., Kim, J., Hong, P. H., Ko, M. J. & Hong, S. W. (2023). Properties and applications of self-healing polymeric materials: a review. *Polymers*, *15*(22), 4408.
[13] Webber, M. J. & Tibbitt, M. W. (2022). Dynamic and reconfigurable materials from reversible network interactions. *Nature Reviews Materials*, *7*(7), 541–556.
[14] Roh, S., Nam, Y., Nguyen, M. T. N., Han, J. H. & Lee, J. S. (2024). Dynamic covalent bond-based polymer chains operating reversibly with temperature changes. *Molecules*, *29*(14), 3261.
[15] Briou, B., Améduri, B. & Boutevin, B. (2021). Trends in the Diels–Alder reaction in polymer chemistry. *Chemical Society Reviews*, *50*(19), 11055–11097.
[16] Utrera-Barrios, S., Verdejo, R., López-Manchado, M. Á. & Santana, M. H. (2022). The final frontier of sustainable materials: Current developments in self-healing elastomers. *International Journal of Molecular Sciences*, *23*(9), 4757.
[17] Gopinath, S., Adarsh, N. N., Radhakrishnan Nair, P. & Mathew, S. (2023). Recent trends in thermo-responsive elastomeric shape memory polymer nanocomposites. *Polymer Composites*, *44*(8), 4433–4458.
[18] Abdou-Sabet, S., Puydak, R. C. & Rader, C. P. (1996). Dynamically vulcanized thermoplastic elastomers. *Rubber Chemistry and Technology*, *69*(3), 476–494.
[19] Du, H., Marin Angel, J., Basak, S., Lai, T. Y. & Cavicchi, K. A. (2021). Cross-linked Poly (Octadecyl Acrylate)/Polybutadiene Shape Memory Polymer Blends Prepared by Simultaneous Free Radical Cross-linking, Grafting and Polymerization of Octadecyl Acrylate/Polybutadiene Blends. *Macromolecular Rapid Communications*, *42*(11), 2100072.

[20] Basak, S. & Cavicchi, K. A. (2023). Structure–property relationships of shape memory, semicrystalline polymers fabricated by in situ polymerization and crosslinking of octadecyl acrylate/polybutadiene blends. *Macromolecular Rapid Communications, 44*(1), 2200404.

[21] Zafeer, M. K. & Bhat, K. S. (2024). Chemical approaches for fabrication of self-healing polymers. *Discover Applied Sciences, 6*(7), 373.

[22] Basak, S. & Bandyopadhyay, A. (2022). Styrene-butadiene-styrene-based shape memory polymers: Evolution and the current state of art. *Polymers for Advanced Technologies, 33*(7), 2091–2112.

[23] Alimardani, M. & Abbassi-Sourki, F. (2015). New and emerging applications of carboxylated styrene butadiene rubber latex in polymer composites and blends: Review from structure to future prospective. *Journal of Composite Materials, 49*(10), 1267–1282.

[24] Yang, L., Wu, M., Yang, X., Lin, B., Fu, L. & Xu, C. (2022). Healable, recyclable, and adhesive rubber composites equipped with ester linkages, zinc ionic bonds, and hydrogen bonds. *Composites: Part A Applied Science and Manufacturing, 155*, 106816.

[25] Xu, C., Nie, J., Wu, W., Fu, L. & Lin, B. (2019). Design of self-healable supramolecular hybrid network based on carboxylated styrene butadiene rubber and nano-chitosan. *Carbohydrate Polymers, 205*, 410–419.

[26] Kuang, X., Liu, G., Dong, X. & Wang, D. (2016). Enhancement of mechanical and self-healing performance in multiwall carbon nanotube/rubber composites via Diels–Alder bonding. *Macromolecular Materials and Engineering, 301*(5), 535–541.

[27] Santana, M. H., Huete, M., Lameda, P., Araujo, J., Verdejo, R. & López-Manchado, M. A. (2018). Design of a new generation of sustainable SBR compounds with good trade-off between mechanical properties and self-healing ability. *European Polymer Journal, 106*, 273–283.

[28] Cruz-Morales, J. A., Gutiérrez-Flores, C., Zárate-Saldaña, D., Burelo, M., García-Ortega, H. & Gutiérrez, S. (2023). Synthetic polyisoprene rubber as a mimic of natural rubber: recent advances on synthesis, nanocomposites, and applications. *Polymers, 15*(20), 4074.

[29] Grange, J. (2018). Functionalization of Polyisoprene: Toward The Mimic of Natural Rubber (Doctoral dissertation, Université de Bordeaux).

[30] Breuillac, A., Caffy, F., Vialon, T. & Nicolaÿ, R. (2020). Functionalization of polyisoprene and polystyrene via reactive processing using azidoformate grafting agents, and its application to the synthesis of dioxaborolane-based polyisoprene vitrimers. *Polymer Chemistry, 11*(40), 6479–6491.

[31] Li, L., Wu, P., Yu, F. & Ma, J. (2022). Double network hydrogels for energy/environmental applications: challenges and opportunities. *Journal of Materials Chemistry A, 10*(17), 9215–9247.

[32] Zechel, S., Hager, M. D., Priemel, T. & Harrington, M. J. (2019). Healing through histidine: Bioinspired pathways to self-healing polymers via imidazole–metal coordination. *Biomimetics, 4*(1), 20.

[33] Zhang, L., Chen, S. & You, Z. (2023). Hybrid cross-linking to construct functional elastomers. *Accounts of Chemical Research, 56*(21), 2907–2920.

[34] Aprem, A. S., Joseph, K. & Thomas, S. (2005). Recent developments in crosslinking of elastomers. *Rubber chemistry and technology, 78*(3), 458–488.

[35] Zheng, Y., Cui, T., Wang, J., Ge, H. & Gui, Z. (2023). Unveiling innovative design of customizable adhesive flexible devices from self-healing ionogels with robust adhesion and sustainability. *Chemical Engineering Journal, 471*, 144617.

[36] Xu, C., Cao, L., Lin, B., Liang, X. & Chen, Y. (2016). Design of self-healing supramolecular rubbers by introducing ionic cross-links into natural rubber via a controlled vulcanization. *ACS Applied Materials and Interfaces, 8*(27), 17728–17737.

[37] Xu, C., Cao, L., Huang, X., Chen, Y., Lin, B. & Fu, L. (2017). Self-healing natural rubber with tailorable mechanical properties based on ionic supramolecular hybrid network. *ACS Applied Materials and Interfaces, 9*(34), 29363–29373.

[38] Imbernon, L., Oikonomou, E. K., Norvez, S. & Leibler, L. (2015). Chemically crosslinked yet reprocessable epoxidized natural rubber via thermo-activated disulfide rearrangements. *Polymer Chemistry, 6*(23), 4271–4278.

[39] Imbernon, L., Oikonomou, E. K., Norvez, S. & Leibler, L. (2015). Chemically crosslinked yet reprocessable epoxidized natural rubber via thermo-activated disulfide rearrangements. *Polymer Chemistry, 6*(23), 4271–4278.

[40] Lan, R., Hu, X. G., Chen, J., Zeng, X., Chen, X., Du, T. . . . Yang, H. (2024). Adaptive liquid crystal polymers based on dynamic bonds: From fundamentals to functionalities. *Responsive Materials, 2*(1), e20230030.

[41] Panferova, L. I., Zubkov, M. O., Kokorekin, V. A., Levin, V. V. & Dilman, A. D. (2021). Using the thiyl radical for aliphatic hydrogen-atom transfer: thiolation of unactivated C– H bonds. *Angewandte Chemie, 133*(6), 2885–2890.

[42] Nguyen, H. N., Lu, L. H. & Huang, C. J. (2024). Aromatic disulfide cross-linkers for self-healable and recyclable acrylic polymer networks. *ACS Applied Polymer Materials, 6*(8), 4615–4624.

[43] Liu, S., Oderinde, O., Hussain, I., Yao, F. & Fu, G. (2018). Dual ionic cross-linked double network hydrogel with self-healing, conductive, and force sensitive properties. *Polymer, 144*, 111–120.

[44] Somseemee, O., Saeoui, P., Schevenels, F. T. & Siriwong, C. (2022). Enhanced interfacial interaction between modified cellulose nanocrystals and epoxidized natural rubber via ultraviolet irradiation. *Scientific Reports, 12*(1), 6682.

[45] Breuillac, A., Caffy, F., Vialon, T. & Nicolaÿ, R. (2020). Functionalization of polyisoprene and polystyrene via reactive processing using azidoformate grafting agents, and its application to the synthesis of dioxaborolane-based polyisoprene vitrimers. *Polymer Chemistry, 11*(40), 6479–6491.

[46] Briou, B., Améduri, B. & Boutevin, B. (2021). Trends in the Diels–Alder reaction in polymer chemistry. *Chemical Society Reviews, 50*(19), 11055–11097.

[47] Tanjung, F. A., Hassan, A. & Hasan, M. (2015). Use of epoxidized natural rubber as a toughening agent in plastics. *Journal of Applied Polymer Science, 132*(29).

[48] Whba, R., Su'ait, M. S., Whba, F., Sahinbay, S., Altin, S. & Ahmad, A. (2024). Intrinsic challenges and strategic approaches for enhancing the potential of natural rubber and its derivatives: A review. *International Journal of Biological Macromolecules*, 133796.

[49] Toiserkani, H. & Rajab-Qurchi, M. (2025). Acrylated epoxidized natural rubber/functionalized organoclay hybrid networks: In-situ production and characterization study. *Polymer Composites, 46*(1), 645–658.

[50] Shundo, A., Yamamoto, S. & Tanaka, K. (2022). Network formation and physical properties of epoxy resins for future practical applications. *Jacs Au, 2*(7), 1522–1542.

[51] Kong, L., Yang, Y., Lin, Z., Huang, B., Liao, L., Wang, Y. & Xu, C. (2024). A ENR-based conductive film integrating electricity-triggered self-healing, damage detection and high sensitivity for flexible sensors. *Chemical Engineering Journal, 479*, 147624.

[52] Hu, W., Wang, C., Fei, F., Wang, R., Wang, J., Tian, H. . . . Zhang, H. (2025). Self-healing epoxidized natural rubber flexible sensors based on hydrogen bonding interactions. *Journal of Materials Chemistry C, 13*(4), 1824–1834.

[53] Ding, H., Wang, Y. & Shen, X. (2023). Dual cross-linked self-healing temperature and pressure sensor based on PAM/ENR/h-BN composite. *Journal of Applied Polymer Science, 140*(26), e53992.

[54] Rahman, M. A., Penco, M., Peroni, I., Ramorino, G., Grande, A. M. & Di Landro, L. (2011). Self-repairing systems based on ionomers and epoxidized natural rubber blends. *ACS Applied Materials and Interfaces, 3*(12), 4865–4874.

[55] Behera, P. K., Kumar, A., Mohanty, S. & Gupta, V. K. (2022). Overview on post-polymerization functionalization of butyl rubber and properties. *Industrial & Engineering Chemistry Research, 61*(46), 16910–16923.

[56] Lewis, S. P. (2004). *Project 1. Synthesis of PIB-silsesquioxane stars via the sol-gel process. Project 2. Solution and aqueous suspension/emulsion polymerization of isobutylene coinitiated by bis (bispentafluorophenylboryl) tetrafluorobenzene* (Doctoral dissertation, The University of Akron).

[57] Kružeлák, J. & Hudec, I. (2018). Vulcanization systems for rubber compounds based on IIR and halogenated IIR: An overview. *Rubber Chemistry and Technology, 91*(1), 167–183.

[58] Cao, R., Zhao, X., Zhao, X., Wu, X., Li, X. & Zhang, L. (2019). Bromination modification of butyl rubber and its structure, properties, and application. *Industrial & Engineering Chemistry Research, 58*(36), 16645–16653.

[59] Porter, A. M. J. (2010). Imidazolium Ionomer Derivatives of Poly (isobutylene-co-isoprene) (Doctoral dissertation).

[60] Li, Q., Yan, F. & Texter, J. (2024). Polymerized and colloidal ionic liquids— syntheses and applications. *Chemical Reviews, 124*(7), 3813–3931.

[61] Das, A., Sallat, A., Böhme, F., Suckow, M., Basu, D., Wießner, S. . . . Heinrich, G. (2015). Ionic modification turns commercial rubber into a self-healing material. *ACS Applied Materials and Interfaces, 7*(37), 20623–20630.

[62] Caruso, M. M., Davis, D. A., Shen, Q., Odom, S. A., Sottos, N. R., White, S. R. & Moore, J. S. (2009). Mechanically-induced chemical changes in polymeric materials. *Chemical reviews, 109*(11), 5755–5798.

[63] Tian, W., Yang, H., Li, H., Wang, S., Jin, H. & Tian, L. (2024). A self-healing polyurethane/imidazole-modified carboxylated graphene oxide composite coating for antifouling and anticorrosion applications. *European Polymer Journal, 219*, 113372.

[64] Li, Q., Wen, C., Yang, J., Zhou, X., Zhu, Y., Zheng, J. . . . Zhang, P. (2022). Zwitterionic biomaterials. *Chemical Reviews, 122*(23), 17073–17154.

[65] Almeida, H. F., Freire, M. G., Fernandes, A. M., Lopes-da-silva, J. A., Morgado, P., Shimizu, K. . . . Coutinho, J. A. (2014). Cation alkyl side chain length and symmetry effects on the surface tension of ionic liquids. *Langmuir, 30*(22), 6408–6418.

[66] Suckow, M., Mordvinkin, A., Roy, M., Singha, N. K., Heinrich, G., Voit, B. . . . Böhme, F. (2017). Tuning the properties and self-healing behavior of ionically modified poly (isobutylene-co-isoprene) rubber. *Macromolecules, 51*(2), 468–479.

[67] Chumnum, K., Kalkornsuraprance, E., Johns, J., Sengloyluan, K. & Nakaramontri, Y. (2021). Combination of self-healing butyl rubber and natural rubber composites for improving the stability. *Polymers, 13*(3), 443.

[68] Luo, G. B., Pang, B., Luo, X. Q., Wang, Y., Zhou, H. & Zhao, L. J. (2023). Brominated butyl rubber anticorrosive coating and its self-healing behaviors. *Chinese Journal of Polymer Science, 41*(2), 297–305.

[69] Shannon, D. P., Cerdan, K., Kim, M., Mecklenburg, M., Su, J., Chen, Y. . . . Hawker, C. J. (2024). Bioinspired metal–ligand networks with enhanced stability and performance: facile preparation of hydroxypyridinone (HOPO)-functionalized materials. *Macromolecules, 57*(24), 11339–11349.

[70] Mphahlele, K., Ray, S. S. & Kolesnikov, A. (2017). Self-healing polymeric composite material design, failure analysis and future outlook: a review. *Polymers, 9*(10), 535.

[71] Ekeocha, J., Ellingford, C., Pan, M., Wemyss, A. M., Bowen, C. & Wan, C. (2021). Challenges and opportunities of self-healing polymers and devices for extreme and hostile environments. *Advances in Materials, 33*(33), 2008052.

[72] Tsarevsky, N. V. & Matyjaszewski, K. (2007). "Green" atom transfer radical polymerization: from process design to preparation of well-defined environmentally friendly polymeric materials. *Chemical reviews, 107*(6), 2270–2299.

[73] Chafiq, M., Chaouiki, A. & Ko, Y. G. (2023). Recent advances in multifunctional reticular framework nanoparticles: a paradigm shift in materials science road to a structured future. *Nano-Micro Letters, 15*(1), 213.

[74] Utrera-Barrios, S., Mas-Giner, I., Manzanares, R. V., Verdejo, R., López-Manchado, M. A. & Santana, M. H. (2024). Recyclability and self-healing capability in reinforced ionic elastomers. *Polymer, 310*, 127468.

[75] Low, D. Y. S., Mintarno, S., Karia, N. R., Manickam, S., Tan, K. W., Khalid, M. . . . Tang, S. Y. (2024). Nano-reinforced self-healing rubbers: A comprehensive review. *Journal of Industrial and Engineering Chemistry*.

[76] Roy, K., Debnath, S. C., Pongwisuthiruchte, A. & Potiyaraj, P. (2021). Review on the conceptual design of self-healable nitrile rubber composites. *ACS omega, 6*(15), 9975–9981.

[77] Sruthi, P. R. & Anas, S. (2020). An overview of synthetic modification of nitrile group in polymers and applications. *Journal of Polymer Science: Polymer Letters Edition, 58*(8), 1039–1061.

[78] Liu, C., Fang, W., Cheng, Q., Qiu, B., Shangguan, Y. & Shi, J. (2025). Revolutionizing elastomer technology: advances in reversible crosslinking, reprocessing, and self-healing applications. *Polymer Reviews, 65*(2), 483–526.

[79] Cheng, Z., Yan, M., Cao, L., Huang, J., Cao, X., Yuan, D. & Chen, Y. (2019). Design of nitrile rubber with high strength and recycling ability based on Fe3+–catechol group coordination. *Industrial & Engineering Chemistry Research, 58*(9), 3912–3920.

[80] Zhang, Z. F., Liu, X. T., Yang, K. & Zhao, S. G. (2019). Design of coordination-crosslinked nitrile rubber with self-healing and reprocessing ability. *Macromolecular Research, 27*, 803–810.

[81] Shang, P., Shao, C., Li, Q. & Wu, C. (2018). Preparation and characterization of high performance NBR/cobalt (II) chloride coordination composites. *Materials Research Express, 5*(2), 025308.

[82] Das, M., Pal, S. & Naskar, K. (2020). Exploring various metal-ligand coordination bond formation in elastomers: Mechanical performance and self-healing behavior. *Express Polymer Letters, 14*(9), 860–880.

[83] Okoro, H. K., Ndlwana, L., Ikhile, M. I., Barnard, T. G. & Ngila, J. C. (2021). Hyperbranched polyethylenimine-modified polyethersulfone (HPEI/PES) and nAg@ HPEI/PES membranes with enhanced ultrafiltration, antibacterial, and antifouling properties. *Heliyon, 7*(9).

[84] Zheng, Y., Li, S., Weng, Z. & Gao, C. (2015). Hyperbranched polymers: advances from synthesis to applications. *Chemical Society Reviews, 44*(12), 4091–4130.

[85] Utrera-Barrios, S., Araujo-Morera, J., De los reyes, L. P., Manzanares, R. V., Verdejo, R., López-Manchado, M. Á. & Santana, M. H. (2020). An effective and sustainable approach for achieving self-healing in nitrile rubber. *European Polymer Journal, 139*, 110032.

[86] Zhong, N. & Post, W. (2015). Self-repair of structural and functional composites with intrinsically self-healing polymer matrices: A review. *Composites: Part A Applied Science and Manufacturing, 69*, 226–239.

[87] Sarkar, S., Banerjee, S. L. & Singha, N. K. (2021). Dual-responsive self-healable carboxylated acrylonitrile butadiene rubber based on dynamic diels–alder "click chemistry" and disulfide metathesis reaction. *Macromolecular Materials and Engineering, 306*(3), 2000626.

[88] De Sarkar, M., Fujii, N., Abe, Y., Kamba, Y. & Sunada, T. (2022). Quest for sustainable curatives for chloroprene rubber: a comprehensive review. *Rubber Chemistry and Technology, 95*(4), 550–574.

[89] Chang, B. P., Gupta, A., Muthuraj, R. & Mekonnen, T. H. (2021). Bioresourced fillers for rubber composite sustainability: current development and future opportunities. *Green Chemistry, 23*(15), 5337–5378.

[90] Xiang, H. P., Rong, M. Z. & Zhang, M. Q. (2016). Self-healing, reshaping, and recycling of vulcanized chloroprene rubber: A case study of multitask cyclic utilization of cross-linked polymer. *ACS Sustainable Chemistry & Engineering, 4*(5), 2715–2724.

[91] Kaur, A., Gautrot, J. E., Cavalli, G., Watson, D., Bickley, A., Akutagawa, K. & Busfield, J. J. (2021). Novel crosslinking system for poly-chloroprene rubber to enable recyclability and introduce self-healing. *Polymers, 13*(19), 3347.

[92] Singha, N. K. & Jana, S. C. (eds.). (2024). Advances in Thermoplastic Elastomers: Challenges and Opportunities.

[93] Dutta, N. K., Bhowmick, A. K. & Choudhury, N. R. (1997). Thermoplastic elastomers. *Plastics Engineering-New York, 41*, 349–380.

[94] Nebouy, M. (2020). Nanostructuration, Reinforcement In The Rubbery State and Flow Properties at High Shear Strain of Thermoplastic Elastomers: Experiments and Modeling (Doctoral dissertation, Université de Lyon).

[95] Chen, Y., Kushner, A. M., Williams, G. A. & Guan, Z. (2012). Multiphase design of autonomic self-healing thermoplastic elastomers. *Nature chemistry, 4*(6), 467–472.

[96] Luo, D., Niu, B., Wang, X. & He, X. (2025). Self-healing thermoplastic elastomers enabled by dynamic ordered microphase crosslinking of random copolymers. *Polymer, 318*, 127964.

[97] Zhang, M. Q. & Rong, M. Z. (2022). Extrinsic and Intrinsic Approaches to Self-Healing Polymers And Polymer Composites. John Wiley & Sons.

[98] Namdari, N., Mohammadian, B., Jafari, P., Mohammadi, R., Sojoudi, H., Ghasemi, H. & Rizvi, R. (2020). Advanced functional surfaces through controlled damage and instabilities. *Materials Horizons, 7*(2), 366–396.

[99] Caruso, M. M., Davis, D. A., Shen, Q., Odom, S. A., Sottos, N. R., White, S. R. & Moore, J. S. (2009). Mechanically-induced chemical changes in polymeric materials. *Chemical Reviews, 109*(11), 5755–5798.

[100] Washiyama, J., Kramer, E. J. & Hui, C. Y. (1993). Fracture mechanisms of polymer interfaces reinforced with block copolymers: transition from chain pullout to crazing. *Macromolecules, 26*(11), 2928–2934.

[101] Aiswarya, S., Awasthi, P. & Banerjee, S. S. (2022). Self-healing thermoplastic elastomeric materials: challenges, opportunities and new approaches. *European Polymer Journal, 181*, 111658.

[102] Liu, C., Fang, W., Cheng, Q., Qiu, B., Shangguan, Y. & Shi, J. (2025). Revolutionizing elastomer technology: advances in reversible crosslinking, reprocessing, and self-healing applications. *Polymer Reviews, 65*(2), 483–526.

[103] Xia, D., Wang, P., Ji, X., Khashab, N. M., Sessler, J. L. & Huang, F. (2020). Functional supramolecular polymeric networks: the marriage of covalent polymers and macrocycle-based host–guest interactions. *Chemical Reviews, 120*(13), 6070–6123.

[104] Ma, X. & Zhao, Y. (2015). Biomedical applications of supramolecular systems based on host–guest interactions. *Chemical reviews, 115*(15), 7794–7839.

[105] Zhang, M., Choi, W., Kim, M., Choi, J., Zang, X., Ren, Y. . . . Lin, Z. (2024). Recent Advances in Environmentally Friendly Dual-crosslinking Polymer Networks. *Angewandte Chemie International Edition, 63*(24), e202318035.

[106] Chen, L., Xu, J., Zhu, M., Zeng, Z., Song, Y., Zhang, Y. . . . Huang, C. (2023). Self-healing polymers through hydrogen-bond cross-linking: synthesis and electronic applications. *Materials Horizons, 10*(10), 4000–4032.

[107] Wolfe, S. V., Tod, D. A. & Rarde,. (1989). Characterization of engineering polymers by dynamic mechanical analysis. *Journal of Macromolecular Science – Chemistry, 26*(1), 249–272.

[108] Zhou, Y., Li, L., Han, Z., Li, Q., He, J. & Wang, Q. (2022). Self-healing polymers for electronics and energy devices. *Chemical Reviews, 123*(2), 558–612.

[109] Gai, Y., Li, H. & Li, Z. (2021). Self-healing functional electronic devices. *Small, 17*(41), 2101383.

[110] Sabet, M. (2024). Unveiling advanced self-healing mechanisms in graphene polymer composites for next-generation applications in aerospace, automotive, and electronics. *Polym-Plast Technol Engineer, 63*(15), 2032–2059.

[111] Pandey, S. K., Mishra, S., Ghosh, S., Rohan, R. & Maji, P. K. (2024). Self-healing polymers for aviation applications and their impact on circular economy. *Polymer Engineering & Science, 64*(3), 951–987.

[112] Bhattacharya, A. B., Chatterjee, T. & Naskar, K. (2020). Automotive applications of thermoplastic vulcanizates. *Journal of Applied Polymer Science*, *137*(27), 49181.

[113] Hel, C. L., Bounor-Legaré, V., Catherin, M., Lucas, A., Thèvenon, A. & Cassagnau, P. (2020). TPV: a new insight on the rubber morphology and mechanic/elastic properties. *Polymers*, *12*(10), 2315.

[114] Babu, R. R. & Naskar, K. (2010). Recent developments on thermoplastic elastomers by dynamic vulcanization. *Advanced Rubber Composites*, 219–247.

[115] Imbernon, L. (2015). Réticulation non-permanente, chimique ou physique, du caoutchouc naturel époxydé: propriétés dynamiques et recyclage (Doctoral dissertation, Université Pierre et Marie Curie-Paris VI).

[116] Faul, C. F. (2014). Ionic self-assembly for functional hierarchical nanostructured materials. *Accounts of Chemical Research*, *47*(12), 3428–3438.

[117] Ariga, K., Li, J., Fei, J., Ji, Q. & Hill, J. P. (2016). Nanoarchitectonics for dynamic functional materials from atomic-/molecular-level manipulation to macroscopic action. *Advances in Materials*, *28*(6), 1251–1286.

[118] Qin, B., Yin, Z., Tang, X., Zhang, S., Wu, Y., Xu, J. F. & Zhang, X. (2020). Supramolecular polymer chemistry: From structural control to functional assembly. *Progress in Polymer Science*, *100*, 101167.

[119] Fischer, H. (2010). Self-repairing material systems—a dream or a reality? *Natural Science*, *2*(8), 873–901.

Chapter 5
Understanding controlled/living polymerization-driven self-healing chemistries: polymerization pathways, structure–property–processing relationships, and recent trends

5.1 Introduction

Self-healing materials have the remarkable capability to autonomously repair physical damages and restore their functionalities without external intervention [1]. This innovative property has attracted significant attention in recent years, driven by the potential to extend the lifespan of materials and devices, thereby promoting sustainability and reducing waste [2]. The concept of self-healing is primarily inspired by natural biological systems, which efficiently repair themselves through complex, multistep processes [3].

Despite their complexity, biological healing processes typically follow three main steps: an immediate biochemical response (such as inflammation), wound closure, and partial or full recovery of functionality. These steps form the foundation for developing self-healing materials and thus are often termed as nature-inspired materials [4]. These biological healing mechanisms have inspired the development of synthetic self-healing materials. By mimicking nature's efficient and effective repair processes, researchers aim to create materials that can autonomously repair themselves after damage, enhancing their durability and longevity. This approach has led to the emergence of a field, often referred to as "nature-inspired materials," where the principles of biological healing are applied to the design of advanced synthetic systems [5–9]. The concept of nature-inspired materials has broad implications across various fields, including materials science, engineering, and biomedical applications. By understanding and replicating the fundamental steps of biological healing, scientists can develop innovative materials with self-repair capabilities, paving the way for more sustainable and resilient technologies [5–9].

In fact, this is quite fascinating, as in biological systems, the healing mechanisms vary between plants and animals but share common steps [4]. For instance, human skin is a familiar example. When minor physical damage such as a small cut occurs, the skin initiates an inflammatory response, triggering specific biochemical processes. This immediate reaction helps to prevent infection and prepares the site for repair. Within hours, the wound begins to close, and over a few days, the skin recovers its original function and properties, such as its sensing capability and high elasticity [10–12].

https://doi.org/10.1515/9783111583716-005

However, synthetic self-healing systems, while inspired by these biological processes, generally involve simpler mechanisms with fewer steps [13]. These systems typically include a sealing phase, characterized by intermolecular diffusion [13], and a chemical or physical repair phase, involving bond rearrangements [14]. Although these processes are simpler, our understanding of self-healing mechanisms in synthetic systems remains limited, posing challenges to the development of effective design principles for materials with desirable self-healing properties [15].

Over the past few decades, a variety of self-healing materials have been developed, including polymers, metals, ceramics, and composites. Each class of materials exhibits unique self-healing mechanisms and applications such as metals can be reused after being melted, but the high energy required to reach their melting points limits their practical use in self-healing applications [16–19]. Consequently, metals with lower melting points, such as liquid metals, are often employed. Examples include liquid metal-based self-healing anode materials for lithium-ion batteries and self-healing liquid metal-based elastomers [20–23].

To counter these disadvantages, ceramics can achieve self-healing through chemical reactions, such as the oxidation of preincorporated healing agents, upon exposure to high temperatures or the atmosphere [24]. This approach allows ceramics to repair themselves without the need for external intervention, for instance, incorporating manganese oxide (MnO) as a doping agent into the ceramic material can significantly accelerate the healing process. When a crack forms, the MnO reacts with the surrounding material to form new compounds that fill and seal the crack, restoring the structural integrity of the ceramic [24–26].

This technique has practical applications in high-stress environments. For example, in the aerospace industry, structural ceramics used in turbine blades of aircraft engines can benefit from this self-healing capability. The high temperatures and mechanical stresses in these environments make durability a critical concern [27, 28]. By enabling the ceramic material to heal itself, the overall lifespan and performance of the turbine blades can be significantly enhanced, reducing maintenance costs and improving safety.

Nevertheless, polymers are particularly advantageous as self-healing materials due to their mechanical flexibility and structural adaptability [4]. The backbone structure, molecular architecture, and functional groups of polymers can be tailored to achieve desired self-healing properties. The self-healing mechanisms in polymers range from the recovery of dynamic bonds to the formation of new chemical bonds from initially isolated chemical precursors (Fig. 5.1) [29, 30]. This versatility allows for the integration of various functionalities within a single polymer, making polymer-based self-healing systems a major research focus. The number of publications on self-healing polymers has grown exponentially, from fewer than 100 per year before 2010 to over 1,000 in 2021 (Fig. 5.1) [4, 29, 30].

Overall, the field of self-healing materials is rapidly advancing, driven by the need for more durable, sustainable, and efficient materials. As research continues to

a

b

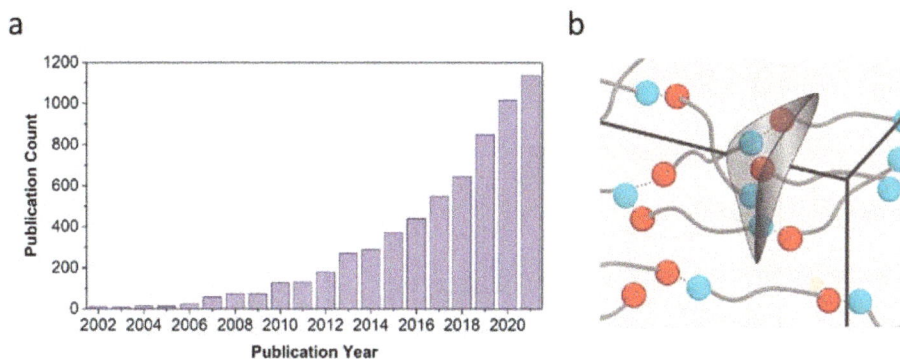

Fig. 5.1: (a) The number of publications per year on topics of self-healing polymers (data from the Web of Science, the search keyword is "self-healing polymer"). (b) Schematic of the material design of intrinsically self-healing polymers, with spheres presenting dynamic reversible bonds, reprinted with permission from [4], copyright reserved American Chemical Society 2022.

uncover the underlying mechanisms of self-healing in synthetic systems, we can expect the development of new materials with enhanced self-repair capabilities, offering promising applications across various industries.

5.2 Performance and service lifetime of self-healing polymers

From an application perspective, the benefits of self-healing materials become clear when examining their performance and service lifetime, compared to traditional materials. Fig. 5.2 illustrates the performance trajectories of original, improved, and self-healing materials over time. Performance is defined by the material's key properties, such as tensile strength, hardness, or corrosion resistance. The service lifetime is the duration during which the material remains above its limit of reliability and in functional condition before replacement is necessary [31].

Curve (a) represents the performance of original materials, and it is inherent that over time, these materials exhibit a steady decline in performance due to wear, damage, and environmental factors. This degradation ultimately limits their service lifetime, necessitating replacement once the material no longer meets the required standards of reliability [32].

However, Curve (b) represents traditionally improved materials. These materials are engineered to provide better durability and resistance to damage compared to original materials. While they offer a slight extension in service lifetime, their performance still gradually diminishes over time, though at a slower rate. Despite these enhancements, traditionally improved materials eventually succumb to wear and damage, requiring replacement after a relatively modest extension of their operational lifespan [31, 32].

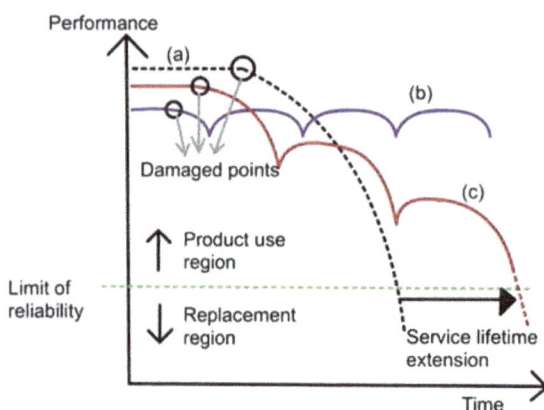

Fig. 5.2: Graph of performance plotted against time for curve (a) (normal material), curve (b) (ideal self-healing material), and curve (c) (self-healing material), reprinted with permission from [31], copyright reserved MDPI 2022.

Interestingly, Curve (c) represents self-healing materials, which exhibit a unique performance trajectory. Upon sustaining damage, the performance of self-healing materials drops initially. However, unlike traditional materials, self-healing materials can autonomously repair themselves, leading to a recovery in performance. This cycle of damage and healing can repeat multiple times. Each time the material heals, its performance increases again, although it may never fully return to its original state. Over time, the performance of self-healing materials gradually decreases due to wear, but at a much slower rate compared to both original and traditionally improved materials. Consequently, the ability of self-healing materials to recover their performance after damage significantly extends their service lifetime [33, 34].

5.2.1 Introduction to the chemistry of self-healing polymers

As we can comprehend, these breakthroughs have been possible due to advancements in polymer chemistry and the ability to fine-tune various functional polymers [35]. Consequently, exploring the chemistries of self-healing polymers remains a pivotal area of interest. To protect or repair damaged or physically cut samples, the design and development of polymeric materials with self-healing or self-repairing characteristics have been extensively studied. The fundamental concept of the self-repairing mechanism in polymers requires a thorough understanding and designing of efficient polymerization reactions. To achieve self-healing properties, various methodologies have been developed. These approaches can be broadly categorized into intrinsic and extrinsic self-healing mechanisms [36, 37].

Intrinsic self-healing involves the incorporation of dynamic covalent or noncovalent bonds within the polymer matrix [38]. These bonds can break and re-form in response to external stimuli, such as temperature, light, or pH changes, allowing the material to heal itself. Examples include reversible Diels–Alder reactions [39], disulfide linkages [40], and hydrogen bonding interactions [41], all of which enable the material to autonomously repair damage at the molecular level.

On the other hand, extrinsic self-healing, on the other hand, relies on the incorporation of healing agents, such as microcapsules or vascular networks, within the polymer matrix [42]. When the material is damaged, these capsules or networks release the healing agents, which then migrate to the damaged area and initiate a repair reaction, often involving polymerization or crosslinking processes. This method mimics biological healing by supplying the necessary components to the site of damage [43, 44].

Nevertheless, recent advancements in polymer chemistry have facilitated the development of multifunctional self-healing polymers that combine both intrinsic and extrinsic healing mechanisms [45]. These hybrid systems offer enhanced healing efficiency and can be tailored to specific applications, ranging from aerospace and automotive industries to biomedical devices and consumer products [46].

Thus, the design and synthesis of self-healing polymers involve a "tethered" approach, integrating principles of chemistry, materials science, and engineering. By using and understanding these scientific advancements, researchers continue to develop innovative materials with improved durability, sustainability, and performance, paving the way for a new generation of smart materials that can autonomously maintain and extend their functional lifespan.

5.2.2 The need for understanding controlled/living polymerization-driven self-healing chemistries

As Fig. 5.1 suggests, neither the concept of self-healing nor self-healing polymers is new. Over the years, there has been a growing body of literature exploring these concepts in depth. Numerous reviews have emerged, highlighting the diverse approaches and advancements in self-healing polymer technologies [47–50]. This chapter covers a wide range of topics, including the fundamental principles of self-healing mechanisms [47–50], the development of various types of self-healing polymers [50–55], and their applications across different fields [55–60].

Self-healing polymers have been the subject of extensive research due to their potential to autonomously repair damage and extend the lifespan of materials. Early studies laid the groundwork by exploring the basic principles and feasibility of self-healing [60–65]. Subsequent reviews have delved into more specific aspects, such as the chemical and physical mechanisms underlying self-healing [15], the incorporation of healing agents [1, 11], and the integration of self-healing properties into different polymer matrices [66–68].

Recent reviews have also focused on advancements in material design and fabrication techniques, highlighting innovations such as the development of advanced polymerization methods, the use of multifunctional additives, and the creation of novel polymer architectures [15, 17, 27, 31].

Although these reviews comprehensively explore and provide a state-of-the-art research on self-healing materials, there is a notable gap in reviews specifically focusing on polymerization techniques. Bielawski and colleagues highlighted the underlying chemistries used in self-healing materials by categorizing them into chemical reactions applicable to autonomous and nonautonomous systems [69]. Fig. 5.3 illustrates self-healing polymers categorized by their underlying chemistry.

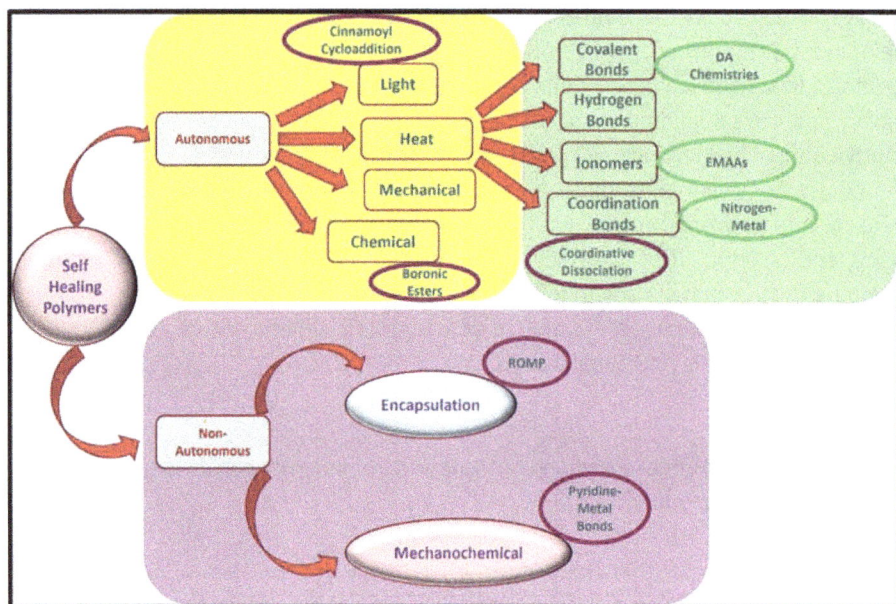

Fig. 5.3: Self-healing polymers and underlying chemical mechanisms and representative chemistries.

Since Bielawski's work, there has been a notable increase in the development of new chemistries for self-healing applications. A particularly promising area is controlled/living polymerization (CLP) [70], which has become increasingly important in designing self-healing polymers and exploring innovative self-healing strategies. CLP techniques have demonstrated significant potential in creating polymers with specific properties that enhance their self-healing capabilities. These methods offer precise control over the polymer's structure and functionality, leading to important advancements in self-healing technologies [70].

Although there has been a notable surge in the synthesis techniques surrounding CLP and significant attention has been directed toward using these advanced polymerization methods to develop innovative self-healing polymers, there is still a lack of comprehensive reviews addressing this burgeoning area of interest. Despite the rapid advancements and growing focus on applying CLP techniques to create polymers with self-healing capabilities, no existing review thoroughly covers the current state of this niche. The absence of such a review leaves a gap in the literature, making it challenging for researchers to access a detailed synthesis of recent progress, trends, and developments in this field. A comprehensive review would be valuable in summarizing and analyzing these advancements, providing insights into how CLP methods are being utilized to enhance self-healing properties and identifying future directions for research and application.

Thus, motivated by this gap in the literature, we aim to introduce a thorough and comprehensive chapter to address it. Our goal is to provide an in-depth analysis of the current state of CLP techniques, highlighting the latest developments and their implications for self-healing materials. This chapter will examine and summarize how CLP methods contribute to the creation of self-healing polymers. We will explore the fundamental principles of CLP, recent innovations, and how these advancements are being utilized to engineer polymers with enhanced self-healing properties.

Additionally, the chapter would delve into how the structural properties of polymers developed through CLP techniques affect their performance in self-healing applications. By understanding the relationship between polymer structure and functionality, the chapter would elucidate how these properties lead to practical applications in various fields, such as materials science and engineering. In summary, this chapter would fill a crucial void by synthesizing recent research on CLP and its impact on self-healing polymers, offering valuable insights into how advancements in polymerization techniques are shaping the future of self-healing materials and their potential applications.

The chapter is systematically organized, beginning with an introduction to the concept of self-healing polymers. We then progress to a detailed examination and evaluation of self-healing chemistries within polymer science. This section covers various chemical mechanisms involved in self-healing processes and provides an overview of CLP, including its general mechanisms and applications.

Following this, we identify and discuss recent advancements and emerging trends in this field. This includes a thorough analysis of how these advancements impact the structure–property relationships of self-healing polymers. We also explore the applications of these materials, illustrating their practical relevance and potential uses. By addressing these aspects, the chapter aims to update the field on the latest developments and clarify how different polymerization techniques affect the properties and effectiveness of self-healing materials. We anticipate that this comprehensive chapter will provide valuable insights into how recent innovations influence self-healing polymers, contributing to a deeper understanding of their capabilities and applications.

5.3 Overview of self-healing polymers from its core chemistry perspective: synthetic polymer chemistry and generational chemical advancements

The field of self-healing polymers has undergone significant transformations since its inception in the 1970s [71]. Initial studies focused on chain interdiffusion mechanisms, but advancements over the decades have led to sophisticated approaches involving dynamic bonds, intrinsic and extrinsic healing mechanisms, and vascular networks [71].

In the 1970s, Malinskii et al. presented pioneering work on polymer self-healing, specifically in poly(vinyl acetate) (PVAc). Subsequent studies by Jud et al. and Wool et al. expanded the understanding of healing cracks in poly(methyl methacrylate) (PMMA), polystyrene (PS), and hydroxy-terminated polybutadiene (HTPB) [71, 72]. These early efforts relied on the concept of chain interdiffusion, a well-known phenomenon in polymers that occurs at temperatures slightly above the glass transition temperature (T_g) of the material [71]. Ellul et al. introduced the concept of self-adhesion in butyl rubber (IIR), emphasizing the importance of good contact between surfaces to ensure effective healing. However, it was Dry et al.'s work in the early 1990s on cement and epoxy resins that laid the groundwork for autonomic self-healing as we know it today [72].

The definitive push for self-healing polymeric materials came with the famous study by White et al. [73]. This work introduced the use of a healing agent (dicyclopentadiene, DCPD) embedded in microcapsules and a platinum catalyst dispersed in an epoxy resin. When damage occurred, the healing agent was released and polymerized upon encountering the catalyst, effectively sealing the crack. This method achieved efficiencies of up to 75% in fracture toughness recovery [73]. Building on this foundation, Keller et al. applied similar methodologies to elastomers, specifically poly(dimethylsiloxane) (PDMS). They utilized a two-microcapsule system containing a high molecular weight resin of PDMS, functionalized with vinyl groups, and a platinum catalyst in one capsule, and an initiator and PDMS copolymer in the other. This approach achieved up to 120% recovery in tear strength, showcasing the potential of external agent-based polymerization for self-healing [62].

The development of self-healing polymers has evolved significantly over the years and can be classified into distinct generations based on their underlying mechanisms and technological advancements (Tab. 5.1). It is also important to note that self-healing materials often mimic mechanisms found in living organisms, such as plants and human skin, to repair or restore damage. These materials are frequently inspired by natural processes as discussed earlier, and thus to ensure the success of self-healing, van der Zwaag defined three key concepts: (a) localization, (b) temporality, and (c) mobility. These concepts essentially describe the robustness of a self-healing process (Fig. 5.4) [74].

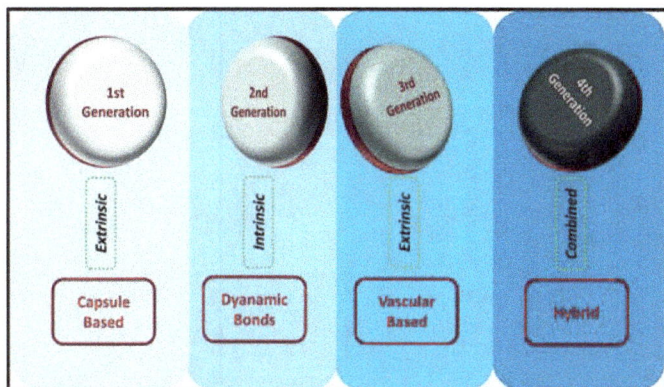

Fig. 5.4: Generations of self-healing materials according to the healing mechanism involved.

The development of self-healing materials hinges on several key approaches (Fig. 5.4), which essentially shape up the niche of synthetic polymer chemistry. They are generally classified as capsule-based (extrinsic), intrinsic, vascular-based, and even a combined approach. Prior to discussing the approaches, it is critical that an understanding is developed regarding the three fundamental key concepts [1–5]. Localization refers to the position and scale of damage in the material, which can be superficial (such as scratches, microcracks, or cuts), deep (including the propagation of surface damage, fiber debonding, or delamination), or at the molecular scale (involving the breakage of the material network) [71, 72]. The self-healing capability is significantly influenced by the localization and scale of damage. While the ultimate goal is to develop a universal protocol for healing all scales of damage, specific protocols are often designed for particular types based on the intended application. Temporality addresses the time gap between the damage event and its repair, aiming to minimize healing time. Enhancing the mobility of the material is a key strategy to reduce this time, as mobility promotes the diffusion of the healing agent to the damaged area and facilitates the re-formation of broken bonds. If the mobility of the agent is inadequate, it will either fail to reach the damage site or do so too slowly, thus reducing the efficiency of the self-healing process. Herein, the approaches come into play. The mechanism categorizes self-healing materials into two main families: extrinsic and intrinsic [12]. Extrinsic self-healing materials rely on an external agent, typically dispersed in the form of capsules or vascular systems, which are released upon damage, to seal the defect without specifically interacting with the matrix. In contrast, intrinsic self-healing materials have reversible bonds that can be restored after a damage event. Extrinsic systems are widely used in thermosets, such as epoxy resins, while intrinsic mechanisms are prevalent in elastomers, including silicones, polyurethanes, and general-purpose rubbers [18–20].

Table 5.1: Generational chemical advancements in self-healing polymers.

First generation	Second generation	Third generation	Fourth generation
–Focus on incorporating healing agents and catalysts into polymer matrices	– Introduction of dynamic bonds (covalent and noncovalent) in self-healing polymers	– Integration of vascular networks into polymer matrices marked this era	– Emerging strategies build on previous approaches while introducing innovative mechanisms
– Self-healing systems classified into:	– Chen et al. pioneered Diels–Alder chemistry with:	– Initially explored by Dry et al., but definitive implementation was delayed due to complexity	– Extrinsic mechanisms applied in:
– Single-capsules	– Multi-furan and multi-maleimide monomers	– Toohey et al. [75] used vascular networks to:	– Coatings
– Capsule/dispersed catalysts	– Achieved 57% recovery in fracture toughness tests at elevated temperatures	– Confine healing agents in epoxy resin coatings	– Instrument panels
– Phase-separated droplets/capsules		– Achieved over 40% healing efficiency	– Sponges
– Double-capsules	– Cordier et al. advanced the field by [75]:	– Supported up to seven healing cycles	– Intrinsic mechanisms show versatility in:
– All-in-one capsules	– Designing supramolecular networks in elastomers using hydrogen bonds	– Challenges encountered in elastomers, with limited success in incorporating robust vascular networks	– General-purpose elastomers
– Exploration of various encapsulation techniques, including:	– Enabled self-healing at room temperature		– Nanogenerators
– In situ polymerization	– Demonstrated the time dependence of the healing process		– Sensors
– Sol–gel reactions			– Conductive elastomers
– Interface polymerization			– Tires
– Emulsion techniques			

The current use of the terms "intrinsic" and "extrinsic," in the context of self-healing materials, differs from their traditional definitions in chemistry. Intrinsic properties are independent of size, shape, and quantity (e.g., density, refractive index), whereas extrinsic properties are dependent on these factors (e.g., weight, volume). Despite these differences, the classification of self-healing materials based on their mechanisms is widely accepted and useful for understanding their development and applications [71].

The development of these advanced self-healing strategies is closely tied to fundamental advances in polymer chemistry and polymerization processes. CLP systems, such as Reversible-Deactivation Radical Polymerization (RDRP), have played a crucial role in this evolution. RDRP techniques offer increased versatility, precise control over polymerization conditions, and the ability to create well-defined polymer structures with specific properties [70]. This control has enabled researchers to design polymers with tailored self-healing capabilities [77]. The success of RDRP is evident from the growing body of literature and its diverse applications. In 2020, Boyer and colleagues reported a comprehensive review that illustrates the cumulative number of publications on RDRP, including both articles and reviews. This chapter highlights the two most commonly used techniques: Atom Transfer Radical Polymerization (ATRP) and Reversible Addition-Fragmentation Chain Transfer (RAFT) Polymerization. The increasing interest in these techniques is also reflected in the patent literature [70].

Since the development of effective RDRP techniques in the 1990s, the field has experienced substantial growth, with over 30,000 publications and approximately 4,000 patents related to RDRP [70]. Initially focused on the development of methods, RDRP has evolved to support a wide range of applications, including therapeutic areas such as nanomedicine, as well as nanotechnology and materials science. This broad applicability demonstrates the significant impact of RDRP on advancing polymer technology and expanding its potential uses, particularly in synthesizing self-healing polymers, which will be explored further in the subsequent sections [70].

5.4 Significance of controlled/living polymerization in solving the challenges of self-healing

A range of effective strategies for CLP have been established, each with its unique reaction mechanisms, reagents, and conditions. Despite these differences, all strategies share a fundamental capability: they enable the simultaneous growth of all polymer chains. This uniform chain growth is achieved under specific conditions [78]. Initially, all polymer chains begin their growth at the onset of polymerization. During this process, the proportion of chains that undergo termination remains relatively small compared to the total number of chains that are either dormant or active [78]. Furthermore, a rapid equilibrium between the dormant and active states of the chains is maintained with the assistance of control agents (X).

The effectiveness of these reactions is significantly influenced by the careful selection of polymerization conditions, initiators (P_1-X), and control agents (X). The concentration of dormant polymer chains (P_n-X) is typically much higher than that of the active propagating radicals (usually with the ratio of $[P_n$-X]/[P_n•] exceeding 100,000) [70]. This high concentration of dormant chains plays a crucial role in reducing the occurrence of irreversible termination between active propagating radicals [70]. As a result, the polymerization process can exhibit features similar to those of living polymerization, where the polymer chains grow in a controlled and predictable manner.

For instance, one of the key benefits of RDRP is the stability and orthogonality of the control agent, which enables continuous polymerization in a manner similar to living systems. This stability allows for precise and controlled polymerization processes, resulting in well-defined polymer structures. In RDRP, the polymer molecular weight increases linearly with the extent of monomer conversion. This linear relationship is achieved by carefully adjusting the stoichiometry of reagents, including the initial concentrations of control agents and monomers, or by precisely managing the extent of monomer conversion [79–81].

For example, a recent study from the Gennaro group showed that the rate of an ATRP reaction was found to be independent of the total concentration of soluble copper species. Instead, the rate of polymerization was predominantly influenced by the ratio of CuI to CuII, which remained constant throughout the course of the polymerization process [79]. However, minor deviations in kinetic behavior were noted at high levels of monomer conversion, particularly at elevated CuI/CuII ratios and at the lowest total concentrations of CuI plus CuII. These deviations are attributable to the inherent radical termination reactions that occur during polymerization. Such reactions lead to a permanent transformation of a portion of CuI into CuII, a phenomenon known as the persistent radical effect (PRE) [79, 80]. This transformation alters the [CuI]/[CuII] ratio over time, which in turn affects the rate of polymerization, generally resulting in a slowdown. Nonetheless, the reaction kinetics associated with RDRP reveal that these polymerization systems are particularly adept at correlating polymer molecular weight with monomer conversion. This effective correlation sheds a fundamental strength of RDRP, making it a hallmark of successful polymer synthesis using this technique.

Secondly, the ability of RDRP to retain chain-end functionality throughout the polymerization process offers a significant advantage. This feature enables the extension of polymer chains by adding more monomers, and thus chain extension allows for the creation of complex polymer structures, such as block copolymers, star copolymers, and brush copolymers [82, 83]. Additionally, RDRP is effective in producing polymers with narrow chain length distributions. When the process of starting polymerization and the rate at which dormant chains transition to active chains are faster than the polymer growth rate, the distribution of polymer chain lengths becomes more uniform. This distribution often resembles a Poisson distribution, indicating consistency in chain lengths, reflecting the precision of the RDRP process [84, 85].

Thus, with this backdrop, we understand that controlled or living polymerization techniques are essential for addressing the practical challenges in the development and application of self-healing materials. These methods provide a high level of precision in polymer synthesis, which is crucial for creating materials with specific properties needed for effective self-healing. For example, CLP allows for accurate control over polymer molecular weight and structure, as we know. This precision is important for tailoring the mechanical properties of self-healing materials to meet specific requirements [86]. For instance, in applications like structural composites or protective coatings, where mechanical strength and flexibility are important, controlled polymerization ensures that materials can be engineered to achieve the desired balance of rigidity and elasticity.

Moreover, these polymerization techniques also enable the precise incorporation of functional groups into the polymer chain, making it important for developing materials with specific self-healing mechanisms. For instance, dynamic covalent bonds or responsive chemical groups can be integrated to enhance the material's ability to repair itself after damage [87]. This ability to customize the polymer chemistry improves the efficiency of self-healing materials in practical applications such as adhesives or coatings [88]. Similarly, CLP also supports the creation of materials with multiple properties by allowing the incorporation of various monomers and functional groups. This capability is valuable for developing self-healing materials that also have additional attributes, such as thermal stability or electrical conductivity. Such multifunctional materials are beneficial in applications ranging from electronic devices to energy storage systems [87 89].

With this in mind, it is interesting to explore the current trends and advancements in RDRP techniques that are revolutionizing the self-healing polymer niche. The next section will examine these developments in greater detail. While early RDRP systems primarily relied on thermal activation at elevated temperatures, recent innovations have focused on conducting these reactions under a variety of conditions to expand their applicability, especially for temperature-sensitive bioapplications [70, 90–91]. Modern RDRP techniques now utilize a diverse array of initiating systems, including those based on electrochemical potentials, mechanical force, enzymatic systems, and light activation. Each of these initiating methods plays a significant role in the development of self-healing polymers [90,91]. By enabling polymerization under more versatile and mild conditions, these advancements highlight how self-healing polymers can be designed and applied, particularly in fields requiring delicate and precise material properties.

5.5 Current strategies and scope in controlled/living polymerization focusing on developing self-healing polymers

5.5.1 Heat-mediated controlled/living polymerization

Polymerization carried out via heat-assisted CLP is one of the most extensively researched themes in the field of self-healing chemistries. This technique involves the use of thermal energy to initiate and control the polymerization process, allowing for precise manipulation of polymer structures and properties. Heat-assisted CLP has been widely studied because it offers several advantages, including the ability to achieve high levels of control over molecular weight distribution and polymer architecture [92, 93].

Researchers have focused on optimizing the conditions under which heat-assisted CLP is performed, investigating factors such as temperature ranges, reaction times, and the types of monomers and catalysts used. These studies aim to enhance the efficiency and effectiveness of self-healing polymers, making them more reliable and versatile for various applications [93].

For example, Singha and coworkers developed a self-healing material, PFMA-co-PBMA, using tailor-made copolymers of furfuryl methacrylate (FMA) and butyl methacrylate (BMA). This material was synthesized through RAFT polymerization by heating the mixture at 80 °C under a nitrogen atmosphere for 16 h. It was further functionalized via click chemistry, incorporating Diels–Alder (DA) and retro-Diels–Alder (rDA) reactions, enabling healing through cross-linking and de-cross-linking at elevated temperatures [43]. The cross-linked DA polymer exhibited significant self-healing properties. When heated at 130 °C in toluene, the polymer became soluble due to the cleavage of the DA product via the rDA reaction, a key feature for self-healing (Fig. 5.5).

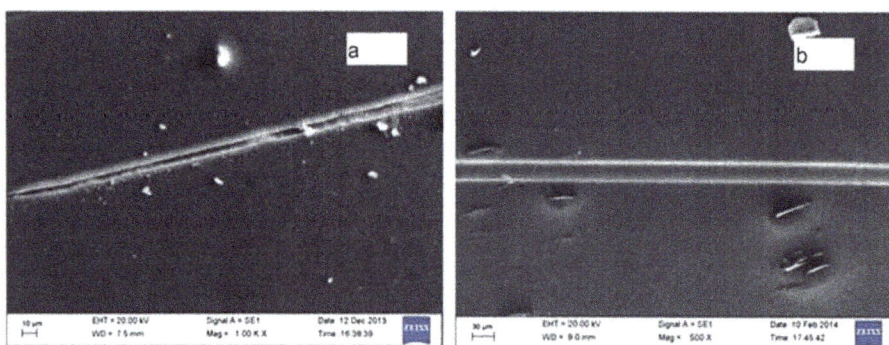

Fig. 5.5: Scanning electron microscopy (SEM) images illustrate the self-healing experiments conducted on the copolymer/BM DA adduct. In these experiments, the material was heated at 80 °C for 2 h and then for 4 h. After each heating period, the temperature was maintained at 50 °C for 24 h, reprinted with permission from [94], copyright reserved Elsevier 2015.

Moreover, during this process, cross-linking and de-cross-linking reactions facilitated the repair of damaged areas, restoring the material's integrity, and thus proving to be a robust shape self-healing polymer design [43].

More advanced self-healing systems are being explored, such as self-healing polystyrene (PS) filled with glycidyl methacrylate (GMA)-loaded poly (melamine-formaldehyde) (PMF) microcapsules [95]. This approach utilizes RAFT polymerization at 70 °C, with S,S'-bis(α,α'-dimethylacetic acid) trithiocarbonate as the RAFT agent, to produce living PS. GMA, a low-viscosity liquid at ambient temperature, with a boiling point of 189 °C, contains reactive carbon–carbon double bonds. Unlike highly volatile styrene, GMA is less volatile, making it easier to encapsulate and enabling it to function both as a solvent and a reactive monomer, enhancing its self-healing potential [95].

Nevertheless, recently, there has been a shift from heat-mediated CLP) techniques to light-mediated methods, driven by advances in additive manufacturing. Traditional thermally-mediated RDRP systems are less suited for photo-induced 3D printing. However, notable developments in fast and oxygen-tolerant photoATRP and photoRAFT polymerization systems show promise for 3D printing applications [96]. These advancements allow for precise control over polymerization under mild conditions, making them compatible with the intricate requirements of 3D printing. The ability to initiate and control polymerization with light enables the fabrication of complex, detailed structures, paving the way for innovative applications in additive manufacturing and advanced materials design [96], as explored in the subsequent section.

5.5.2 Light-mediated controlled/living polymerization

As discussed, beyond thermal initiation, photochemical activation has become the most frequently used method for activating and controlling RDRP. Numerous photochemical strategies have been employed for radical generation in RDRP, including conventional type I and type II photoinitiation systems under UV irradiation [97], which have historically been used for radical generation in uncontrolled radical polymerization processes. Following radical generation, RDRP can proceed via the standard mechanisms, previously outlined [97].

In recent years, more advanced strategies have been developed that enable the external control of the activation–deactivation equilibrium through the application of light, leading to what is often referred to as photo-controlled RDRP [98]. These systems provide spatiotemporal control over the polymerization process, allowing it to be precisely stopped and started on demand in a spatially resolved manner. This capability not only enhances the versatility and precision of RDRP but also opens up new possibilities for its application in fields requiring high control over the polymerization process, such as 3D printing and other advanced manufacturing techniques. The ability to manipulate polymerization with light has thus become a powerful tool in the development of novel materials with tailored properties [97–99].

For example, Boyer et al. developed and implemented a fast and oxygen-tolerant PET-RAFT polymerization system within a DLP (Digital Light Processing) 3D printing framework. In their system, erythrosin B (EB) and triethanolamine (TEtOHA) were employed as a photocatalyst and co-catalyst, respectively, to promote efficient photoactivation of RAFT agents [100]. The process begins with EB in its excited state oxidizing TEtOHA, which results in the formation of a reduced photocatalyst ($PC^{\cdot-}$) and a corresponding tertiary amine radical cation ($NR_3^{\cdot+}$). Subsequently, $PC^{\cdot-}$ reduces the RAFT agent to revert to its ground state, while the reduced RAFT agent undergoes homolytic cleavage to produce a thiocarbonylthio-stabilized anion species and reactive carbon-centered radicals ($P_n^{\cdot}$). These radicals are then capable of adding across monomer vinyl bonds, facilitating network growth, controlled by degenerative chain transfer between active and dormant RAFT agent-capped chains, thus allowing for the dynamic reformation of bonds, enabling the material to recover from damage effectively [100].

Recent advancements in polymer chemistry have led to the development of several innovative strategies designed to enhance and fine-tune the self-healing properties of materials [100–103]. These advancements are driven by enabling more precise control over the thermal and mechanical characteristics of polymers, which, in turn, significantly improves their self-healing abilities and broadens their range of applications [96]. Among these advancements are photo-initiated RAFT polymerization, NTI-RAFT (Norrish Type I RAFT) polymerization, photoATRP (photo-induced Atom Transfer Radical Polymerization), and RPC-RAFT (photo-induced radical-promoted cationic) polymerization [96]. For example, although RPC-RAFT polymerization is not technically an RDRP technique due to the use of cationic, rather than radical, chain carriers, its degenerative chain transfer process exhibits similarities to other Degenerative Transfer Radical Polymerization (DTRP) methods [96]. Consequently, RPC-RAFT polymerization shares some of the benefits associated with RDRP techniques during network synthesis, and is included here for completeness [96].

The RPC-RAFT polymerization process combines aspects of RAFT polymerization with radical-promoted cationic polymerization. This method utilizes several key components: a cationically polymerizable monomer, an onium salt, a photo-initiator, and a RAFT agent. The inclusion of these components provides flexibility in adjusting reaction conditions, which helps in producing polymers with optimized thermal and mechanical properties. This versatility makes RPC-RAFT polymerization particularly valuable for creating materials capable of self-healing under various environmental conditions. As a result, such materials are well-suited for demanding applications like automotive parts, infrastructure components, and other fields where durability and long-term performance are critical [103].

Similarly, photoATRP harnesses light to initiate and control the ATRP process, offering a high degree of spatial and temporal precision. This approach allows for the creation of polymers with intricate, multifunctional structures that can display enhanced self-healing properties [104]. PhotoATRP is especially useful for applications that require detailed and responsive materials, such as in the 3D printing of customizable parts, the development of tissue engineering scaffolds, and the design of responsive surfaces [96].

5.5.3 Mechanochemistry-mediated controlled/living polymerization

The development of mechanochemistry has significantly influenced the field of polymer chemistry, particularly in the creation of self-healing polymers. Mechanochemistry explores how mechanical forces can trigger chemical reactions, and this principle has inspired the development of polymer systems that can be activated by mechanical stimuli [105]. One notable application is the use of mechanoinitiation in RAFT and ATRP, which has advanced the capabilities of self-healing polymers. For instance, Esser-Kahn and colleagues have demonstrated an innovative approach by employing commercially available barium titanate ($BaTiO_3$) piezoelectric nanoparticles, in conjunction with mechanical force, specifically ultrasonication, to induce ATRP. In their study, the piezoelectric nanoparticles convert the ligated Cu^{2+} deactivator complexes into Cu^+ activators, when subjected to ultrasonic agitation. This process enables the ATRP of n-butyl acrylate (nBA) using an alkyl bromide initiator [105]. The continuous ultrasonic agitation in this method facilitates time-dependent polymer growth, providing a dynamic route for controlled polymerization [105].

It is speculated that the integration of mechano-initiated polymerization techniques into self-healing polymers offers several advantages. Firstly, it enhances self-healing efficiency by allowing precise control over the polymerization process, which can lead to more effective and rapid repair of damaged areas. Secondly, these techniques are adaptable to various environments [107]. For instance, the use of piezoelectric nanoparticles and ultrasonic agitation can be tailored to different settings, making the materials useful across a range of applications from automotive to biomedical fields [108].

Additionally, the dynamic and on-demand nature of mechano-initiation means that the self-healing process can be activated, as needed. This capability allows for real-time repair, making the materials particularly valuable in applications requiring immediate restoration of functionality. Moreover, the use of commercially available materials and established polymerization techniques, such as ATRP and RAFT, simplifies the integration of self-healing polymers into existing manufacturing processes, accelerating their adoption in various industries [107–109].

5.5.4 Electrochemistry-mediated controlled/living polymerization

In 2011, Matyjaszewski and colleagues introduced an innovative approach to controlling ATRP through the application of electric current, a method termed electrochemically mediated ATRP, or eATRP. This technique utilized a cathodic current to reduce the deactivator, a ligated copper complex consisting of Cu^{2+}-Br_2 and Me_6TREN, into a Cu^+-Br/Me_6TREN species [110]. The reduced Cu^+ species then facilitates the ATRP process by promoting the regular halide transfer from dormant halide-capped polymer chains [110]. The key advantage of eATRP lies in its ability to provide precise control over the polymerization process. By manipulating the applied electric potential, current, and total charge, researchers can finely tune the polymerization rate. This capability allows for dynamic adjustments to the polymerization conditions, enabling the process to be paused and restarted, as needed, through periodic switching of the applied potential. This level of control enhances the versatility of ATRP, allowing for the synthesis of polymers with highly controlled molecular weights and architectures [110].

Building on the success of eATRP, recent research has explored the extension of electrochemical control to other RDRP techniques, including RAFT and Nitroxide-Mediated Polymerization (NMP). The electrochemically mediated version of RAFT, termed eRAFT, and the electrochemically mediated NMP, have been investigated in several recent publications. These methods leverage electrochemical stimuli to modulate the activation and deactivation processes involved in RAFT and NMP, providing similar benefits of enhanced control and flexibility in polymer synthesis [110–114].

The precise regulation of polymerization via external electric currents not only boosts the efficiency and adaptability of polymer synthesis but also aligns with the broader objective of creating more sophisticated and functional polymer materials [114,115]. A notable example of this innovation is the work by Matyjaszewski and colleagues, who have pioneered the development of star-shaped block copolymers with a maltotriose core. These star-shaped copolymers feature arms made from acrylates with distinct soft and hard segments: poly(n-butyl acrylate) (PnBA) and poly(tert-butyl acrylate) (PtBA) (Fig. 5.6) [115]. The synthesis of these copolymers was accomplished through a streamlined, two-step process employing a "core-first" approach. This process utilized a simplified electrochemically mediated atom transfer radical polymerization (seATRP) that required only a minimal amount of copper catalyst – just 25 parts per million (ppm). The precise control afforded by this method allowed for the successful formation of both the soft and hard segments, resulting in well-defined macromolecules with narrow molecular weight distributions ($M_w/M_n = 1.13$ for the PnBA homopolymer and $M_w/M_n = 1.17$ for the PnBA-b-PtBA block copolymer) (Fig. 5.6) [110].

The self-healing capabilities of these star-shaped copolymers were demonstrated experimentally, showing remarkable performance [115]. After damage, the self-repair process was observed to occur within minutes, significantly reducing the extent of damage. Within 30 min, the scratch was completely repaired and no further changes in the film's morphology were observed over a prolonged period of two days. These self-healing properties were achieved at room temperature and without requiring any external stimuli, comprehending the effectiveness of the self-healing mechanism [115].

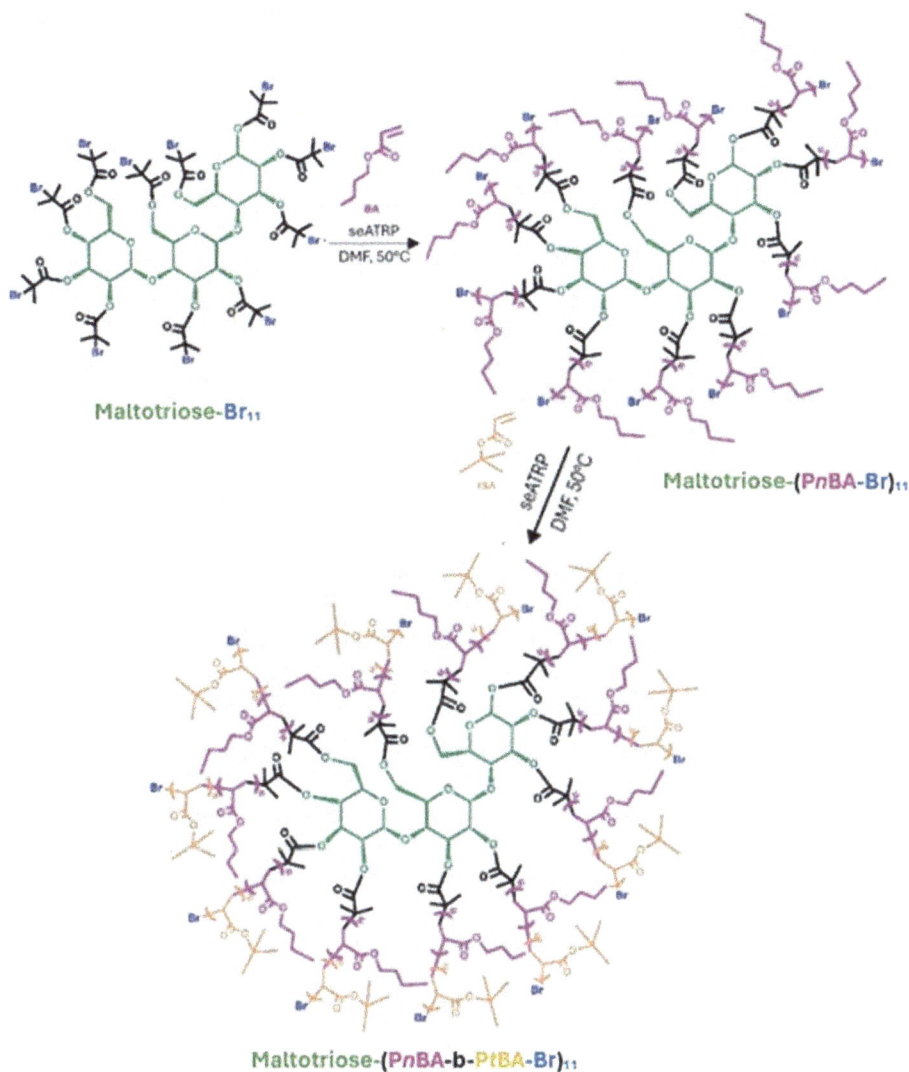

Fig. 5.6: Synthetic route for the preparation of star-like macromolecules with maltotriose core and acrylate arms, reprinted with permission from [115], copyright reserved Elsevier 2022.

One key advantage of these advanced CLP techniques is their ability to offer precise control over polymerization processes. This control allows for the production of polymers with specific, well-defined properties that are often superior to those achieved using conventional methods. For example, self-healing polymers developed through advanced CLP methods typically exhibit better-defined mechanical properties, more efficient self-repair mechanisms, and improved overall performance compared to those made using traditional polymerization techniques.

5.6 Advances in controlled/living polymerization and how these influence the structure–property–self-healing relationships

5.6.1 Living anionic polymerization

Living anionic polymerization is a well-established synthetic method used to develop advanced self-healing polymers. This technique is characterized by its ability to maintain control over the polymerization process due to the absence of chain transfer and chain termination reactions [116–118]. In living anionic polymerization, once the polymerization begins, the growth of the polymer chains continues without interruption. The absence of chain transfer means that the polymer chains do not exchange segments with other chains, which maintains the integrity of the polymer structure. Similarly, the lack of chain termination ensures that the chains continue to grow, leading to well-defined polymers [116].

As we discussed, one key feature of this method is that the rate of chain initiation, the process by which new polymer chains are started, is typically faster than the rate of chain propagation, the process by which existing chains grow. This leads to a situation where the number of active polymer chains remains constant throughout the polymerization process. As a result, the polymers produced have a narrow distribution of molecular weights, meaning that most of the polymer chains have similar lengths [116]. Polymers with narrow molecular weight distributions often exhibit improved performance in applications requiring consistency and reliability. For instance, they can be used in coatings, adhesives, and other materials where uniform properties are essential. In the same context, Barthel et al. employed the living anionic polymerization technique to create precisely engineered block copolymers composed of poly(ethylene oxide)-b-poly(furfuryl glycidyl ether) (PEO-b-PFGE) for use as self-healing materials [116]. In their approach, thermo-reversible bonds were formed within polymeric sheets by a reaction between furan groups in the PFGE segments and a bifunctional maleimide, which served as a cross-linking agent. This reaction, known as the DA reaction, facilitated the self-healing properties of the resulting block copolymers (Fig. 5.7a) [116].

The self-healing capabilities of these polymers were thoroughly evaluated using various techniques, including depth-sensing indentation, profilometry, and differential scanning calorimetry (DSC). The analysis showed that the block copolymers could effectively heal scratches measuring 1.7 micrometers in depth, 0.22 millimeters in width, and 3.5 millimeters in length, even after multiple healing cycles (Fig. 5.7) [116]. After the crosslinking process, the block copolymer films exhibited a hardness of 0.6 GPa and a stiffness of 5.13 GPa, highlighting their viscoelastic nature (Fig. 5.7) [116]. To better understand the mechanical characteristics of PFGE-based films, a detailed examination was conducted before and after cross-linking with BMA. Initially, the PFGE55 films (lacking PEO segments) were in a liquid state at room temperature, preventing direct indentation measurements. In contrast, the PEO 330-b-PFGE 20 block copolymer demonstrated measurable hardness (0.038 GPa) and stiffness (0.46 GPa), attributed to the incorporation of PEO moieties (Fig. 5.7) [116].

Interestingly, following cross-linking with BMA, the PFGE 55 films exhibited a significant increase in mechanical properties, achieving a hardness of 0.6 GPa and a stiffness of 5.13 GPa – values that surpass those of traditional hard polymers such as PMMA (0.32 GPa) and polystyrene (0.34 GPa). This enhancement is primarily due to the high degree of cross-linking facilitated by the furan groups in PFGE 55, which participate in Diels–Alder reactions with bifunctional maleimide, resulting in a robust network structure (Fig. 5.7). This more robust network improves the material's resistance to mechanical stress and deformation, which is particularly critical for self-healing materials. The enhanced resistance enables the material to maintain its performance and functionality even after sustaining damage. Consequently, this resilience supports effective self-healing mechanisms, as a stronger and stiffer material is less likely to undergo irreversible damage and can better recover its properties through self-healing processes [116].

In contrast, the PEO 330-b-PFGE 20 block copolymer, with a lower PFGE weight fraction (17 wt%), exhibited reduced hardness (0.071 GPa) and stiffness (0.91 GPa) after cross-linking (Fig. 5.7). However, these values are approximately twice as high as those in the non-cross-linked state, due to the partial cross-linking of the PFGE segments within the copolymer matrix, contributing to improved mechanical properties compared to the uncross-linked material [116]. The mechanical behavior of the block copolymer films displayed viscoelastic properties, with PEO segments acting as a softening agent and increased hardness being observed at higher displacements. This viscoelasticity is vital for self-healing applications, allowing effective deformation and shape recovery (Fig. 5.7) [116].

From Fig. 5.7, it is evident that following initial cross-linking, the films underwent thermal treatment at 155 °C for 3 h, resulting in a color change from red to brown and a decrease in modulus (Ei) to 1.76 GPa and hardness to 0.26 GPa for PFGE 55. This reduction was attributed to incomplete cross-linking and partial degradation. Subsequent re-cross-linking at 65 °C for 14 h partially restored the mechanical properties of PFGE 55, with the modulus increasing to 4.51 GPa and hardness rising to 0.45 GPa, although still lower than after the initial cross-linking, due to incomplete Diels–Alder reactions (Fig. 5.7) [116].

Overall, while the viscoelastic behavior of these block copolymer films supports effective self-healing through deformation and recovery, factors like thermal treatment and cross-linking extent significantly influence the stability and effectiveness of the self-healing process (Fig. 5.7). Optimizing these conditions remains a challenge for enhancing self-healing capabilities and overall material performance.

Anionic polymerization can be highly advantageous for specific applications, particularly in fields where precision and reliability are crucial. For instance, electrical treeing, a phenomenon where electrical discharges cause degradation within polymer insulation, is a major issue impacting the longevity and reliability of electrical power systems and power electronics [119]. Electrical treeing accelerates the breakdown of

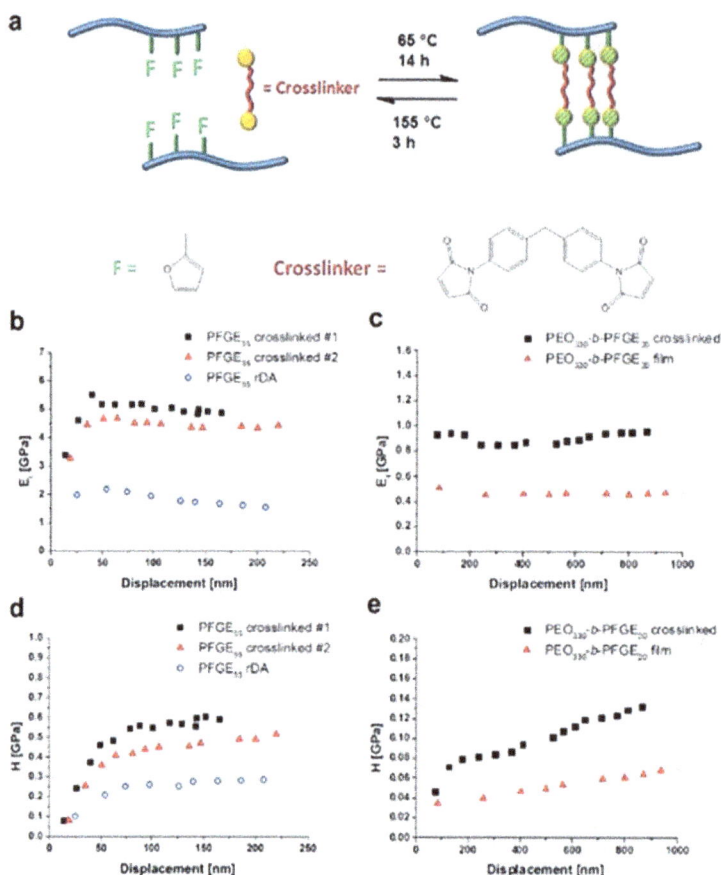

Fig. 5.7: (a) A diagram illustrating the cross-linking process of PFGE using a bifunctional maleimide cross-linker, BMA. (b) and (d) Depth-sensing indentation measurements for PFGE55 films, showing their properties before and after cross-linking. (c) and (e) Depth-sensing indentation measurements for PEO330-b-PFGE20 films, also highlighting their characteristics before and after cross-linking, reprinted with permission from [116], copyright reserved Wiley 2013.

insulation materials, leading to short circuits and equipment failure. This is especially problematic in bulk thermoset polymers, which are commonly used as electrical insulators, because self-healing mechanisms for such electrical damage are rarely reported in these materials [119].

Recent advancements by Xie et al. [119] introduced a novel method for self-healing electrical damage in thermoset polymers, specifically epoxy, using a microcapsule approach that employs anionic polymerization. Their technique involves modifying the epoxy matrix with 2-ethyl-4-methylimidazole, which serves as an initiator for anionic polymerization [119]. This modification enables the in situ polymerization of healing agents contained within microcapsules upon electrical damage (Fig 5.8). The key advantage of this approach is its compatibility with moderate temperature ranges, aligning with the operational conditions of many polymer insulation materials in electrical power systems. This design eliminates the need for additional stimuli, such as extreme temperatures or chemical activators, which are often necessary in other self-healing systems [119].

It is interesting to note that the healing agent is formulated by combining diglycidyl ether of bisphenol F (DGEBF) with polypropylene glycol diglycidyl ether (PPGDGE) [119]. Different ratios of these components were tested to optimize viscosity and dielectric strength. A higher viscosity can hinder microcapsule preparation and healing agent flow, while a lower viscosity can reduce dielectric strength post-healing (Fig 5.8) [119]. Additionally, fewer microcapsules are needed for electrical damage compared to mechanical damage, as the healing agent naturally migrates toward the path of electrical damage due to its higher permittivity and conductivity [119].

Although living anionic polymerization results in increasingly strong structural polymers, its counterpart, cationic polymerization, has also been used to develop smart materials. Specifically, these polymerization techniques can be combined with various chemical methods to create more controlled and customized polymer systems designed for very specific applications, which the next section will discuss.

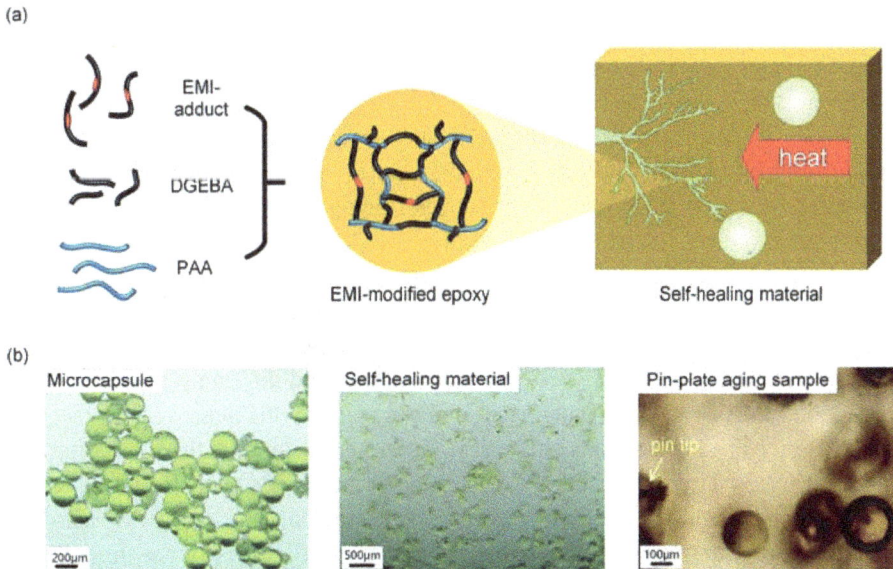

Fig. 5.8: (a) Schematic representation of the synthesis and mechanism of the self-healing material. Microcapsules were added into the EMI-modified epoxy to give the material self-healing ability. (b) Optical photographs of the microcapsule, self-healing material, and pin–plate aging sample. In the right picture, the microcapsules that seemed to be located close to each other are actually in the upper and lower layers, and are separated, reprinted with permission from [119], copyright reserved Royal Society of Chemistry 2020.

5.6.2 Living cationic polymerization

Early research on cationic polymerization included the development of self-healing epoxy systems [120-122]. For instance, Xiao et al. (in 2009) synthesized a two-component healing agent composed of epoxy- and $((C_2H_5)_2O \cdot BF_3)$-loaded microcapsules and used it to fabricate self-healing epoxy composites. The curing of the epoxy healing agent, catalyzed by $(C_2H_5)_2O \cdot BF_3$, is a type of cationic chain polymerization, characterized by its rapid reaction at ambient temperature and low catalyst concentration [120]. Experimental results demonstrated that cracks in composites containing this healing system could be quickly re-bonded with satisfactory healing efficiency. For example, with 5 wt% epoxy- and 1 wt% $(C_2H_5)_2O \cdot BF_3$-loaded capsules, approximately 80% recovery of impact strength was achieved within 30 min at 20 °C. Because the healing capsules were effective at low content, the mechanical properties of the matrix were largely retained [120].

The impact of healing agent capsules on the mechanical properties of composites is noteworthy [120]. Typically, incorporating fillers into polymers alters their intrinsic mechanical properties, so it was essential that the healing agent capsules did not adversely affect these properties. Mechanical tests indicated that adding either $(C_2H_5)_2O \cdot BF_3$- or epoxy-loaded capsules minimally impacted the tensile and flexural properties of the epoxy composites, aside from a slight decrease in flexural strength and modulus for epoxy-loaded capsules. This suggests strong interfacial interactions between the capsules and the matrix, allowing the microcapsules to carry some load under static testing conditions [120].

Further analysis revealed that the healing efficiency of the epoxy composites was significantly influenced by the size and distribution of the microcapsules. Smaller, uniformly distributed microcapsules enhanced healing performance due to increased surface area and consistent healing agent release [120]. Additionally, the wall thickness of the microcapsules was critical in controlling the release rate; thinner walls facilitated faster release, leading to quicker healing of damaged areas. The chemical composition of the healing agent is vital for optimizing self-healing properties. Specifically, modifying the molecular structure of the epoxy and the catalyst has enhanced compatibility between the healing agent and the matrix. This improvement leads to more effective healing reactions and overall better performance of the composites.

Similarly, Hondred et al. developed self-healing thermosets from the cationic polymerization of tung oil, styrene, and divinylbenzene. Traditional cationic polymerization typically utilizes moisture-sensitive Lewis acids as initiators, which can be problematic in environments where controlling moisture is challenging. To address this issue, Hondred et al. explored the use of rare earth triflates as initiators, which offer a new approach to polymerization where moisture sensitivity is not a significant concern [121].

Recent developments have also focused on developing innovative microcapsule-based strategies. Notably, Guo et al. pioneered the first single SiO_2 microcapsule self-healing system based on UV-induced cationic polymerization for potential application in aerospace coatings (Fig. 5.9I) [122]. The innovative aspect of this system lies in its ability to encapsulate both the epoxy resin and a cationic photoinitiator solution in propylene carbonate into a single robust SiO_2 microcapsule. The resulting SiO_2 microcapsules exhibit properties such as solvent resistance and thermal stability [122]. They are especially special for their strong resistance to thermal cycling in simulated space environments, a critical requirement for aerospace applications. The high repairing efficiency of these microcapsules is attributed to several factors: the high load of healing species within the microcapsule, the accurate stoichiometric ratio of the encapsulated components, and the UV-triggered healing chemistry (Fig. 5.9I) [122]. The encapsulated healing agents in the microcapsules ensure that, upon damage, the healing process can be initiated effectively by UV light. This UV-induced cationic polymerization mechanism is advantageous because it is not sensitive to oxygen, a common issue in many self-healing systems (Fig. 5.9I). This oxygen insensitivity facilitates the self-healing application of the single SiO_2 microcapsule in aerospace coatings, where exposure to varying atmospheric conditions is inevitable.

It is crucial to note that for any practical application, the robustness of a self-healing system's resistance to long-term thermal cycling is extremely significant, especially for aerospace self-healing coatings that undergo the harsh thermal-cycled space environment. To evaluate this, Guo et al. simulated the high/low temperatures of thermal cycling and the low atmospheric pressure, characteristic of space environments, to assess the self-healing performance of their single SiO_2 microcapsules (Fig. 5.9II). In their study, Guo et al. compared the pristine SiO_2 microcapsules with those that had undergone thermal cycling, within a temperature range of -50 to 110 °C for five days. The pristine microcapsules (Fig. 5.9IIa) maintained their spherical morphology even after thermal cycling (Fig. 5.9IId), indicating their robustness under such extreme conditions [122].

To further investigate the self-healing capability, both the pristine and thermally cycled microcapsules were embedded into amine-cured epoxy coatings. These coatings were then deliberately scribed with a razor blade to introduce scratches. SEM images (Fig. 5.9II b and 5.9II e) showed that the scribe region had a width of about 20–40 µm. When exposed to UV irradiation for 30 min, the scratches in both the pristine microcapsule-based self-healing coatings and the thermally cycled microcapsule-based self-healing coatings were almost completely filled (Fig. 5.9II c and 5.9II f). This healing effect is attributed to the simultaneous flow of epoxy resin and the cationic photoinitiator released from the ruptured microcapsules toward the cracks, followed by UV-triggered curing to achieve the self-healing effect. The results confirm that the single SiO_2 microcapsule containing epoxy resin and a cationic photoinitiator is highly resistant to the extremely harsh thermal cycling environment. This robustness is essential for aerospace applications, where materials are subjected to rapid and extreme temperature fluctuations [122].

Moreover, the efficient self-healing process observed in these microcapsules under UV irradiation is particularly advantageous for aerospace coatings. In the space environment, abundant UV radiation is readily available, providing a natural trigger for the self-healing mechanism. This capability ensures that any damage sustained by the coatings can be autonomously repaired, thereby extending the lifespan and maintaining the integrity of aerospace components.

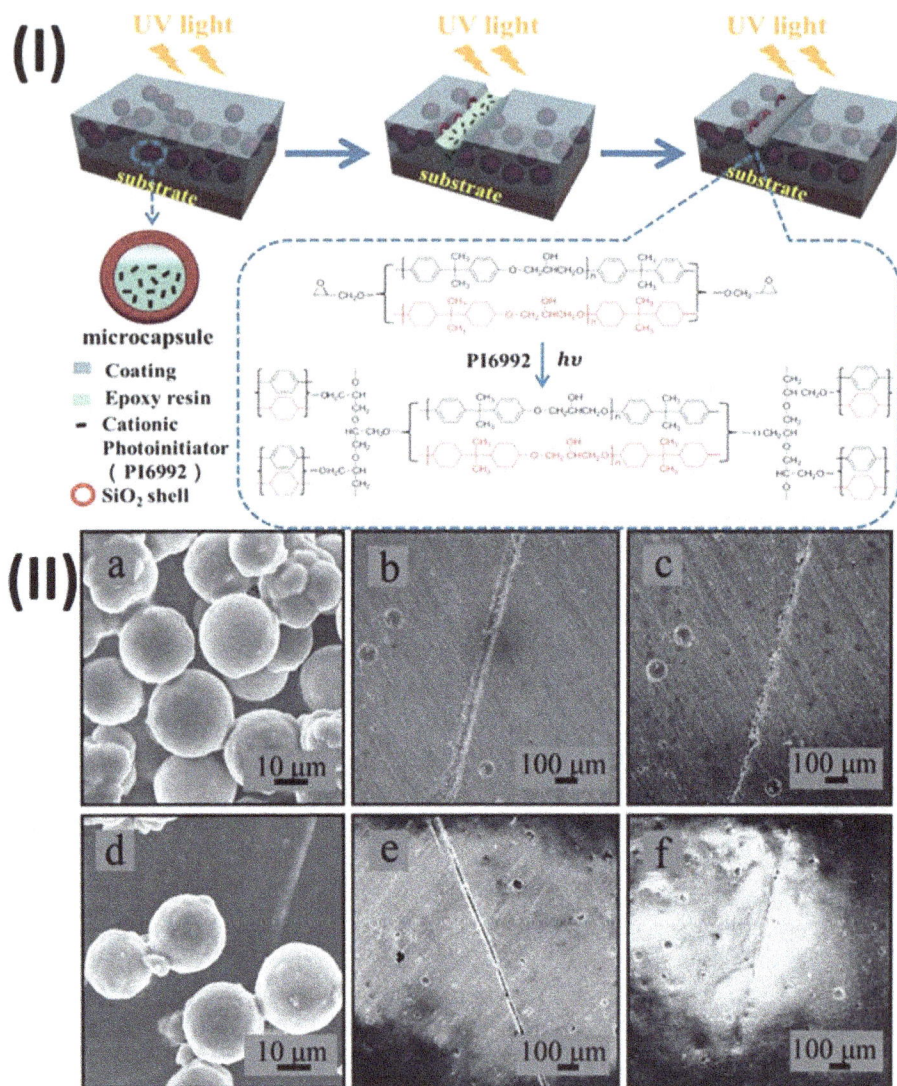

Fig. 5.9: (I) Illustration of UV-triggered self-healing of SiO_2 microcapsules embedded into epoxy resin coating, reprinted with permission from [122], copyright reserved American Chemical Society 2016. (II) SEM images of the pristine microcapsules (a), the microcapsules after thermal cycling in the temperature range from − 50 to 110 °C for 5 d, the scribe region of the self-healing coating containing the pristine microcapsules before (b) and after (c) UV irradiation for 30 min, and the scribe region of the self-healing coating containing the thermally cycled microcapsules before (e) and after (f) UV irradiation for 30 min, reprinted with permission from [122], copyright reserved American Chemical Society 2016.

Combining chemistries is now an emerging trend that allows for more versatile polymers with robust physical properties, providing tunable features for tailored engineered applications. This approach integrates different polymerization and cross-linking strategies to create advanced materials with enhanced functionalities [30]. For example, Binder et al. reported hyperbranched azide- and alkyne-functionalized poly(isobutylene)s (PIBs) with molecular weights of 25,200 to 35,400 g/mol, suitable for room-temperature cross-linking via copper-catalyzed alkyne–azide "click" cycloaddition (CuAAC). The low glass transition temperature (Tg) of these PIBs enhances their molecular mobility, crucial for effective crosslinking [123]. The PIBs were synthesized using inimer-type living carbocationic polymerization (LCCP), initiated from a multifunctional inimer to form hyperbranched structures [123]. The living chain ends were quenched with 3-(bromopropoxy)benzene (BPB) or trimethyl(3-phenoxy-1-propynyl)silane (TMPPS), introducing reactive end groups that facilitate further functionalization [123]. This method allows for tuning the polymer's molecular architecture by adjusting polymerization conditions and end groups, resulting in materials with tailored properties for specific applications.

In addition, other researchers have explored various chemistries for advanced self-healing materials. For instance, hyperbranched azide- and alkyne-modified PIBs have been developed that enable room-temperature self-healing without external stimuli [125–127]. Long et al. synthesized amphiphilic polymers through esterification and ring-opening polymerization, yielding materials with low toxicity and biocompatibility [128-131]. Herbst et al. combined living carbocationic polymerization and azide–alkyne "click" reactions to investigate the self-healing properties of PIBs with hydrogen-bonding moieties, which exhibited effective healing at room temperature after physical contact at fractured surfaces [132].

However, while CLP techniques, such as anionic/cationic polymerization, are indeed useful and can lead to a range of interesting properties that can be further tailored for various applications; they often come with complexities that limit their practical feasibility. The intricate and multistep nature of these reaction processes can make them challenging to implement consistently, particularly in industrial or high-throughput settings. This complexity can restrict their utility as a general-purpose polymerization toolkit. Anionic polymerization, for instance, requires precise control over reaction conditions and purity to avoid issues such as unwanted side reactions or incomplete polymerization. The need for stringent conditions, including an oxygen-free environment and exact temperature control, further complicates its application. These factors can make the process less adaptable to varying production needs or less feasible for scaling up.

In response to these challenges, RDRP technologies are evolving to address these limitations. RDRP techniques offer several advantages over traditional CLP methods. They generally involve simpler and more flexible reaction conditions, which can be easier to control and adapt. RDRP methods, such as ATRP and RAFT polymerization, have shown the ability to handle a wide range of monomers and produce polymers with well-defined structures and properties, often with fewer restrictions on the reaction environment. These advancements make RDRP techniques a more versatile and practical choice for a variety of applications. They offer the ability to synthesize complex polymers with controlled molecular weights and narrow molecular weight distributions while being more tolerant of diverse reaction conditions and impurities. As a result, RDRP is increasingly becoming a preferred method for developing polymers with specific properties tailored to meet the demands of various industrial and research applications.

5.6.3 Reversible addition–fragmentation chain-transfer polymerization

Generational advancements in self-healing polymers have closely followed significant progress in synthetic chemistry [133]. Continuous improvements in polymerization methods have enabled the development of new macromolecules with controlled architectures and precisely placed functionalities essential for effective self-healing. Among these advancements, RAFT polymerization, pioneered by CSIRO researchers in 1998, stands out as a transformative technique in the field of precision polymer development [134].

RAFT polymerization utilizes thiocarbonylthio (TCT) compounds to mediate radical polymerizations, providing a powerful tool for controlling polymer growth and tailoring polymer properties. The RAFT process operates through a mechanism known as degenerate chain transfer, where the propagating radical of a polymer chain is transferred to a TCT compound. This transfer process effectively "quasi-stops" the polymerization, allowing for the precise control of chain length and the subsequent reinitiation of polymer growth. This ability to regulate chain growth is crucial for designing polymers with specific physical and chemical characteristics [135-137]

One notable advantage of RAFT polymerization is its ability to produce polymers with well-defined molecular weights and narrow polydispersities [138–140]. This precision allows for the design of polymers with specific functionalities, such as enhanced self-healing properties, which are critical for applications in demanding environments. For instance, in self-healing polymers, RAFT enables the incorporation of functional groups that respond to external stimuli or facilitate the repair of damage [134–140].

In the context of self-healing polymers, RAFT polymerization has facilitated the development of materials with controlled architectures that enhance their ability to repair themselves. By precisely tuning the polymer structure and functionality, researchers have been able to create polymers that can autonomously restore their integrity after being damaged. This is usually achieved through mechanisms such as reversible cross-linking or the release of healing agents in response to damage.

Researchers have recently synthesized a tailor-made copolymer using furfuryl methacrylate (FMA) and trifluoroethyl methacrylate (TFEMA) through RAFT polymerization. This copolymer's unique properties were further enhanced by modifying the furfuryl groups via the DA reaction with varying molar contents of polyhedral oligomeric silsesquioxane maleimide (POSS-M). This modification resulted in hydrophobic fluorinated copolymers with different POSS content, significantly enhancing their properties [141]. The incorporation of hydrophobic POSS molecules significantly improved the surface hydrophobicity of the modified copolymers. While the parent fluoro copolymer exhibited a water contact angle (WCA) of approximately 101°, the DA-modified polymers achieved a WCA of 135°, demonstrating a substantial increase in hydrophobicity. Additionally, the dynamic covalent furan-POSS-M linkages endowed these copolymers with thermo-reversible behavior, which was crucial for their self-healing capabilities [141].

The developed self-healing hydrophobic POSS-modified DA fluoropolymers present a promising avenue for advanced applications, particularly as specialty paints and coatings. Their enhanced hydrophobicity and self-healing properties make them ideal candidates for protective coatings in various industries, ensuring longevity and durability, while reducing maintenance costs.

In the same context, recent advancements, post 2020, have significantly contributed to the field of RAFT polymerization [142–145]. These architectural advancements and well-defined polymerizations pave the way for achieving elevated goals in shape-memory and self-healing polymers, particularly in response to stimuli. Among various stimuli, the conversion of light into microstructure changes or mechanical motions has garnered particular interest. Another noteworthy research involved the fabrication of a biomass resource-based well-defined polymer nanocomposite, termed CNTs-graft-poly(tetrahydrofurfury methacrylate-co-lauryl acrylate-co-1-vinylimidazole) copolymer (CNTs-g-P(TMA-co-LA-co-VI)), through RAFT polymerization. Subsequent metal–ligand cross linking with Zn^{2+} ions resulted in CNTs-g-P(TMA-co-LA-co-VI)/Zn^{2+} composites, with a CNTs content of 1.1 wt %, exhibiting a maximum stress of 1.68 MPa and an elongation at break of 450% [144].

Significantly, the photothermal conversion of CNTs effectively triggered the association and dissociation process of dynamic metallosupramolecular cross-linked bonds between VI and Zn^{2+}, leading to excellent instantaneous multiple shape-memory and accurate self-healing performance under NIR light or heat. This approach holds promise for the generalized design of self-healing and shape-memory nanocomposites with tunable mechanical properties [144].

While we have discussed traditional RAFT polymerization and its role in promoting superior self-healing polymers, with some past works captured in previous reviews [1, 3, 4, 17, 27, 31, 47, 51, 53], a major question we want to address is how these systems are evolving in recent trends.

Moreover, while traditional RAFT polymerization has been crucial in developing superior self-healing polymers, recent trends indicate a significant shift toward integrating newer methodologies that enhance these systems [146]. Traditional RAFT is known for producing well-defined polymers with controlled molecular weights and architectures, utilizing thiocarbonylthio compounds to mediate radical polymerizations. However, recent advancements, such as Photo Electron/Energy Transfer-Reversible Addition–Fragmentation Chain-Transfer (PET-RAFT) polymerization [147], are addressing its limitations and expanding its applications. PET-RAFT employs photo-induced electron or energy transfer mechanisms to enable polymerization under ambient conditions using visible light, making it more cytocompatible than traditional UV-initiated processes. This shift is particularly advantageous for biomedical applications where UV light can be harmful.

Recent studies have perceived the practical applications of PET-RAFT polymerization, particularly in the development of self-healing hydrogels. Researchers have effectively utilized PET-RAFT to fabricate hydrogels from commercially available monomers, achieving high monomer conversions and efficient cell encapsulation [147]. These hydrogels exhibit the desired rheological and mechanical properties and demonstrate excellent cytocompatibility. Notably, they can be cut and healed by adding further monomer, and irradiating with visible light, even in the presence of mammalian cells. A key hypothesis is that the presence of the chain transfer agent at the end of each polymer chain and the residual end groups within the hydrogel matrix could facilitate further chain propagation even after the formation of the three-dimensional (3D) network. This characteristic would allow the hydrogels to self-heal and repair after being damaged [147].

To test this hypothesis, Rigby et al. performed an experiment using a polyethylene glycol diacrylate (PEGDA) hydrogel with a degree of polymerization (DP) of 50. The hydrogel was cut in half with a scalpel, and the two halves were placed close to each other, but not in full contact. PEGDA monomer (100 µL) was then added to fill the gap between the halves, and the hydrogel was irradiated for 60 min. The resultant healed hydrogel demonstrated the ability to support its own weight when suspended through a needle, with no observable separation between the two glued hydrogel blocks (Fig. 5.10).

The mechanical properties of the healed hydrogels were then evaluated and compared with those of the hydrogels prior to cutting. The PEGDA (DP = 50) hydrogel, prepared with a higher end-group concentration, was chosen for this assessment due to its superior mechanical performance compared to the PEGDA (DP = 100) hydrogel. Following self-healing, the mechanical properties of the hydrogel showed a slight decrease, with the Young's modulus reducing from 360.9 kPa to 273.8 kPa and the stress at break decreasing from 138.2 kPa to 105.0 kPa. Despite this reduction, the healed materials retained good mechanical performance, indicating that the PET-RAFT polymerization process effectively supports the restoration of structural integrity and functionality in the hydrogel [147].

Recent advancements in RAFT polymerization have also led to the development of innovative self-healing materials with improved structural properties. One notable

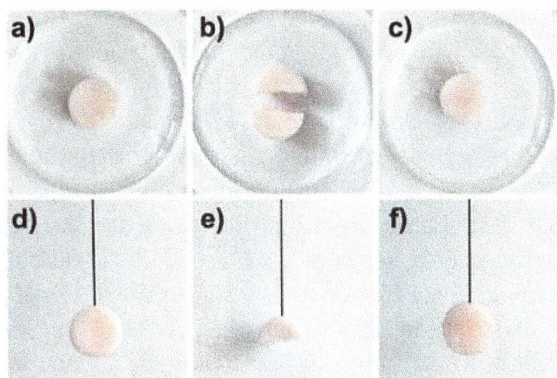

Fig. 5.10: PET-RAFT polymerized hydrogels exhibit self-healing properties, as demonstrated with PEGDA (DP 50) hydrogels. Initially, the PEGDA (DP 50) hydrogel is shown in its intact state (a). After being cut in half (b), the hydrogel is healed through blue light irradiation for 60 min (c). The intact PEGDA (DP 50) hydrogel is capable of being suspended through a needle (d), while the cut hydrogel (e) demonstrates successful healing and the ability to support its own weight (f), reprinted with permission from [147], copyright reserved American Chemical Society 2023.

application is the creation of polymeric coatings, such as nanocrystals/fluorinated polyacrylate containing coumarin derivatives, designed to withstand harsh environmental conditions that can cause cracks and degradation [148]. Researchers synthesized dual-functional cellulose nanocrystals/fluorinated polyacrylate materials using RAFT-assisted Pickering emulsion polymerization [148]. This process showed a gradual increase in droplet size and a decrease in creaming layer as pH increased, with TEM revealing effective anchoring of modified cellulose nanocrystals on latex particles, eventually resulting in excellent water resistance and impressive repeatability in self-healing capabilities [148].

Another significant development involves novel functional polymers such as bottle-brush polymers, characterized by densely grafted side chains [149]. A new method using selective photoactivation in PET-RAFT polymerization was introduced, utilizing a methacrylate monomer with a dodecylthiocarbonothioylthioylthio unit. PET-RAFT polymerization was catalyzed by zinc naphthalocyanine-based moiety (ZnTtBNc) under near-infrared light at room temperature, with a specific chain transfer agent. Grafting poly(methyl acrylate) (PMA) from PDTPEMA via thermal-initiated RAFT polymerization produced well-defined P(DTPEMA-g-PMA) bottle-brush polymers, resulting in films with excellent scratch self-healing properties due to the interlocking nature of the side chains [149]. By grafting poly(methyl acrylate) (PMA) from PDTPEMA via thermal-initiated RAFT polymerization, well-defined P(DTPEMA-g-PMA) bottle-brush polymers were successfully synthesized. The resulting films exhibited excellent scratch self-healing properties due to the interlocking nature of the side chains, highlighting the effectiveness of this advanced polymerization method [149].

Similarly, developments in processing techniques have been increasingly integrated with RAFT polymerization to develop innovative solutions for self-healing materials. These techniques not only enhance the functionality of polymers but also enable the creation of more versatile and practical materials for various applications [150].

For instance, Boyer and colleagues have demonstrated a novel approach where 3D-printed polymers, using a DBTTC (dodecyl thiocarbonothioylthio) RAFT agent, can undergo a welding process (Fig 5.11) [151]. This process is facilitated by reactivating the RAFT species within the material after it has been cut [150]. As illustrated in Fig 5.11c, the cut polymer is exposed to 365 nm UV radiation, which initiates the self-healing process. Remarkably, the material can regain approximately 49 ± 5% of its original tensile strength and 26 ± 2% of its elongation at break within just 5 min of exposure. With prolonged irradiation, the healing efficiency improves further, achieving up to 98 ± 3% of the original tensile strength and 74 ± 3% of the elongation at break within 30 min (Fig. 5.11) [150].

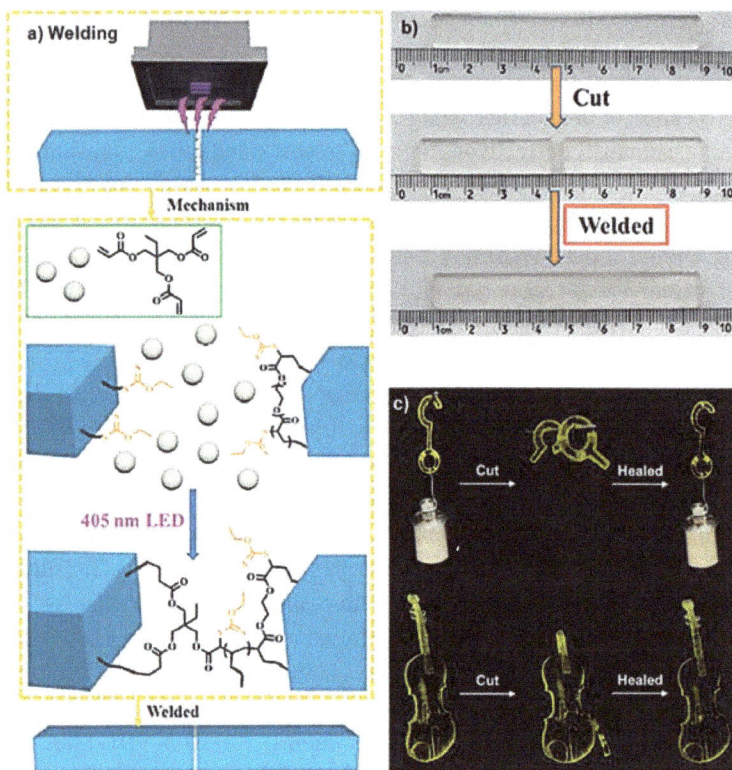

Fig. 5.11: (a) Scheme of the polymer welding process of cut samples. (b) Digital images of a PEGDA film (prepared using a xanthate) before and after welding. (c) Photo-induced self-healing/welding of 3D printed object via RAFT polymerization using a DBTTC, reprinted with permission from [150], copyright reserved American Chemical Society 2023.

In a parallel study, Li, Zhu, and coworkers investigated a different welding process using photo-iniferter mechanisms for 3D-printed polymers. In their method, polymers such as PEGDA (poly(ethylene glycol) diacrylate) or trimethylolpropane triacrylate were painted on the surfaces of cut samples. These samples were then exposed to purple light for 20 min [152]. The presence of the RAFT agent at the cut interface initiated post-polymerization, effectively repairing the cut film. This approach demonstrates that self-healing can be achieved through post-processing of the printed polymers, leveraging the RAFT polymerization's ability to reinitiate polymer growth and repair [152].

Although 3D printing using photopolymerization has been explored for several decades, the integration of RAFT polymerization into 3D printing techniques represents a significant advancement that began in 2019. This approach, often referred to as "living" 3D printing, has expanded the possibilities of additive manufacturing by enabling the creation of materials with unique properties and functionalities that traditional techniques cannot achieve. Initially, RAFT-mediated 3D printing was developed to produce materials that could be reshaped or reformed after their initial fabrication [150]. This capability led to the exploration of different RAFT-based chemical mechanisms, such as photo-iniferter RAFT, PET-RAFT, cationic RAFT, and photo-initiator RAFT. Each method introduced new advantages, allowing for a wide range of applications [150–153]. These advancements demonstrate the potential of combining RAFT polymerization with 3D printing, offering effective material repair and customization. However, there is still room for further development, including exploring different light wavelengths in RAFT processes and creating new RAFT agents and polymer formulations to improve performance and broaden applications [150–153].

Another trend is the development of hybrid polymerization techniques that combine RAFT with other advanced methods to achieve novel functionalities. For instance, integrating RAFT with techniques such as ring-opening metathesis polymerization (ROMP), click chemistry, or other catalytic processes allows for the creation of polymers with complex architectures and multifunctional properties. These hybrid approaches can further enhance the performance of self-healing polymers by introducing additional functionalities or improving material properties [153].

Thus, while traditional RAFT polymerization has laid the groundwork for developing advanced self-healing polymers, recent trends indicate a shift toward integrating new methodologies that offer enhanced capabilities and applications. Techniques like PET–RAFT represent a significant advancement, providing improved control, cytocompatibility, and versatility [154].

Amidst the remarkable advancements in RAFT polymerization, its future implications are rapidly expanding. While the fundamental mechanisms of RAFT are well-established, there remain numerous untapped opportunities to optimize livingness and enhance product quality further. Recent investigations in dispersed media have highlighted the significance of understanding RAFT kinetics to mitigate termination rates, thereby enhancing the quality of block copolymer products. Additionally, there is a burgeoning interest in controlling molecular weight distribution, as it offers avenues to tailor the physical properties of polymeric materials. Various strategies proposed by Fors' and Boyer's groups have demonstrated effective control in this regard [135]. Furthermore, recent studies by Anastasaki and colleagues have explored methods to modulate polymer dispersity by employing high- and low-activity TCT compounds, showing promise in fine-tuning the bulk physical properties of RAFT-derived polymers. Moreover, the impact of initiator-derived radical concentrations on RAFT remains relatively unexplored, presenting opportunities to enhance traditional initiation systems [133]. With the expiration of the original RAFT patents, increased industrial adoption of RAFT can be anticipated. Ongoing efforts to integrate RAFT into existing radical polymerization processes are expected to yield significant benefits.

5.6.4 Atom transfer radical polymerization

Control over radical polymerizations relies on two key principles: rapid initiation to ensure a consistent concentration of growing polymer chains and the persistent radical effect, which keeps most chains dormant, while allowing them to grow through a dynamic equilibrium between dormant species and active radicals. This balance minimizes termination by maintaining a low concentration of active radicals throughout the polymerization process [155]. ATRP fulfills these criteria by utilizing a transition metal catalyst, in combination with a suitable ligand, creating a reversible equilibrium between growing radicals and dormant species. The ligand choice enhances catalyst solubility and modulates this equilibrium. With a low concentration of propagating radicals compared to dormant chains, the proportion of terminated chains is minimal (<5%), resulting in highly functional polymers (>95%) [155, 156].

By varying the initiator, functional end groups can be introduced into linear polymer chains. Typically, the initiator is a low molecular weight organic compound (RX) containing an activated halogen (X), which leaves the fragment (R) at one end of the chain after initiation. The halogen can then be transformed into various functionalities through standard organic procedures [155, 156]. ATRP's radical nature also allows the polymerization of diverse functional monomers, yielding polymers with pendant functional groups that can be further utilized to create block copolymers due to the "living" nature of the process [155, 156].

A particular example of the importance of ATRP in designing self-healing polymers can be seen in the development of self-healing hydrogels for biomedical applications. In this process, the RAFT agent ensures precise control over the polymer chain growth, allowing for the incorporation of specific functional groups that facilitate self-healing [157–161]. This capability is particularly important in biomedical applications where hydrogels are used for drug delivery or tissue engineering. The self-healing properties ensure that the hydrogels can maintain their structural integrity and functionality over time, even after physical damage, thus providing reliable and sustained performance in the body. This example highlights how controlled radical polymerizations like RAFT can create advanced self-healing materials with significant practical benefits [157–161].

A lot of these advancements have been captured in the reviews by Zhu et al. [162], Yang et al. [1], and Wang et al. [163]. However, the recent developments were not covered, and we wanted to focus on the same in this section. For example, nature has been inspiring a lot lately. Ma et al. recently developed a novel approach combining surface-initiated photo-induced electron transfer ATRP and mussel-inspired chemistry for the surface engineering of graphene oxide to create self-healing hydrogels.

In this study, Ma et al. leveraged the natural adhesive properties of mussel-inspired chemistry to modify graphene oxide (GO), enabling it to participate in the ATRP process. This combination resulted in a self-healing hydrogel with enhanced mechanical properties and self-repair capabilities. The graphene oxide was first functionalized with catechol groups, mimicking the adhesive proteins found in mussels. This modification allowed for strong adhesion and interaction within the hydrogel matrix [160].

Understanding the structure–property relationships of polymers is crucial for exploring how GO-type nanomaterials, healing time, and the molecular weight of poly(4-vinylpyridine) (P4VP) affect self-healing hydrogels [160]. Stress–strain tests showed that hybrid-PAA hydrogels with polydopamine exhibited superior strength, toughness, and healing efficiency compared to those without it. Specifically, the self-healing efficiency (f) of nanocomposite hydrogels with polydopamine was 94.9%, while those without ranged from 44.1% to 54.2% [160].

Self-healing efficiencies increased with time, achieving 44.6%, 67.9%, 75.1%, and 94.9% at 2, 4, 6, and 8 h, respectively, using P4VP with a molecular weight of 39,500 g/mol (M_w/M_n = 1.38). This confirmed that polydopamine-based hydrogels could restore 94.9% of their original mechanical strength after 8 h of healing. Higher P4VP molecular weights improved strength and self-healing efficiency, reaching optimal values of approximately 6.94 MPa and 94.9% after 12 h of photocatalysis [160]. Reducing GO content from 1 mg to 0.5 mg decreased strength and healing efficiency to 4.31 MPa and 79.1%, respectively. Load–unload tests indicated effective energy dissipation, with the hydrogels remaining intact after 10 cycles [160].

Additionally, experiments with dyed and unstained gel strips demonstrated excellent mechanical and self-healing properties. When cut and healed together, the strips supported 200 g for about 2 h and did not break, when stretched to 15 cm, indicating their robustness (Fig. 5.12) [160].

Thus, this study highlights the promising potential of integrating GO-type nanomaterials, polydopamine, and P4VP in the development of self-healing hydrogels with enhanced mechanical properties. Moreover, the green methodology, which combines surface-initiated PET–ATRP with eco-friendly mussel-inspired chemistry, offers a novel nanotechnology for the surface modification of nanomaterials. This approach opens up potential applications in tissue engineering, wearable sensors, and adsorp-

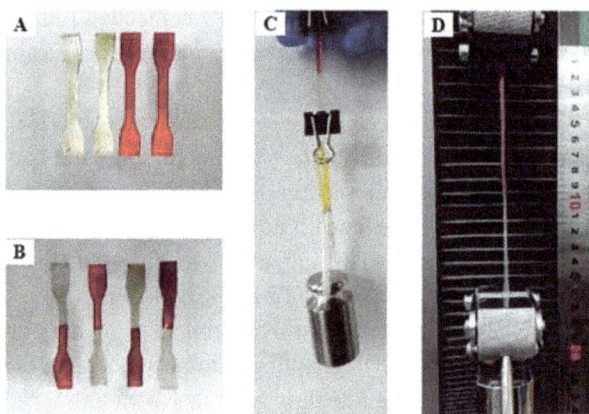

Fig. 5.12: The images of self-healing process with stretched strips: with and without dyed (a), healed stretched strips in different colors (b), healed stretched strips supporting 200 g weight (c) and stretching (d), reprinted with permission from [160], copyright reserved Elsevier 2020.

tion materials. Research has shown that ATRP-based polymerization using mussel chemistry can develop polymer systems for innovative applications in these fields [157, 164, 165].

Future research on self-healing hydrogels could focus on optimizing polymer molecular weights and varying nanomaterial contents to enhance performance. Additionally, studying the long-term stability and durability of these materials under different environmental conditions could facilitate practical applications in biomedicine, soft robotics, and flexible electronics [166, 167]. Emerging strategies in self-healing polymers include Zhao et al.'s work on Van der Waals-driven healing in copolymers with a "lock-and-key" architecture, designed to enable recovery from structural damage. A key challenge in this approach is the nonuniform sequence distribution of copolymers during polymerization, which limits effective site interactions and complicates van der Waals-driven healing [166–168]. To address this, methods for synthesizing lock-and-key copolymers with prescribed sequences have been developed to create architectures conducive to self-healing [166].

Thus, to create architectures conducive to self-healing, methods for synthesizing lock-and-key copolymers with prescribed sequences have been developed. The recovery behavior of three poly(n-butyl acrylate/methyl methacrylate) [P(BA/MMA)] copolymers – alternating (alt), statistical (stat), and gradient (grad) – was evaluated. Despite having similar molecular weights, dispersity, and overall composition, the alt and stat copolymers exhibited a tenfold increase in recovery rate compared to the grad variant, even with similar glass transition temperatures. Small-angle neutron scattering (SANS) indicated that rapid recovery is linked to a uniform microstructure that prevents chain pinning in glassy MMA-rich regions [166].

To assess self-healing capability, the recovery rate of mechanical properties in 0.2 mm-thick films was tested using cut-and-adhere methods at ambient conditions (22 °C). Fig. 5.13 shows the recovery process (Fig. 5.13a) and the extensibility (350% strain) of the alternating copolymer system after 3 h of healing (Fig. 5.13b) [166].

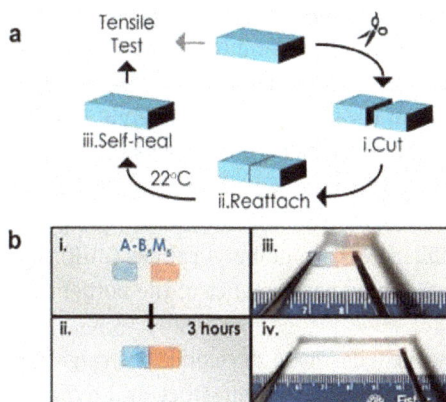

Fig. 5.13: Illustration of self-healing test process by cut and adhere. (b) Photographs demonstrating 350% extensibility of A-B₅M₅ bulk film, 3 h after reattachment. To enhance visibility, films in the photographs were selectively dyed with oil-based blue and red pen ink, reprinted with permission from [166], copyright reserved American Chemical Society 2023.

Recent advances in ATRP-based self-healing materials have shifted focus from traditional applications to functional aspects, particularly with the development of bottle-brush polymers [169–173]. Characterized by densely grafted side chains that extend the polymer backbone, these polymers exhibit significantly reduced chain entanglement compared to linear counterparts. This unique arrangement minimizes inter-chain interactions, which typically lead to entanglements [169–173]. The reduced entanglement in bottlebrush polymers enhances their self-healing capabilities. Lightly cross-linked bottlebrush polymers represent a promising approach for developing su-persoft materials that operate without solvents [169–173]. The extended side chains create a flexible and dynamic network, allowing for efficient chain realignment during the healing process. This property makes them particularly suitable for applications requiring flexibility and rapid recovery from structural damage [174].

However, the extreme softness of bottlebrush polymers makes them prone to flowing, necessitating cross-linking through permanent covalent bonds to maintain structural integrity. Unfortunately, these covalent cross-links can hinder self-healing of mechanical damage. Researchers are exploring various chemical interactions to develop effective self-healing mechanisms in bottlebrush polymers [174–176].

Recent work has focused on a novel physical self-healing mechanism that leverages the unique structure of bottlebrush polymers: the interlocking of long side chains

[174]. By adjusting the length and density of these side chains using ATRP, researchers have developed bottlebrush polymers with remarkable healing efficiency, reaching up to 100%. Molecular dynamics (MD) simulations reveal that the long side chains of adjacent polymers interlock through side-chain interactions. This interlocking is stabilized by van der Waals forces and molecular entanglements, which help maintain the network structure even under stress, and thus this opens up to a number of interesting applications such as in shock absorption and vibration damping. For example, the polymer's peak value of the loss factor (tan δ) reaches 1.6, indicating a high level of energy dissipation. At room temperature, the tan δ value is approximately 0.85 [177], which is substantially higher than the 0.3 threshold, typically considered sufficient for damping materials. The temperature range over which tan δ exceeds 0.3 spans about 80 °C, from 8 °C to 88 °C. This broad temperature span is significantly wider compared to commercial elastomers such as natural rubber (NR), bromobutyl rubber (BIIR), and polyurethane (PU), highlighting the superior damping performance of the bottlebrush polymers [174].

Beyond damping applications, Fan and colleagues have developed a straightforward yet effective strategy that leverages both molecular architecture and dynamic bonds to create dynamically cross-linked bottlebrush polymer materials with integrated self-healing and adhesive properties [178]. As this approach combines ATRP, ROMP, and the "click" thiol-bromo coupling reaction, it presents several advantages over previously reported strategies for producing high-performance bottlebrush polymer materials. Firstly, the bromine (Br) end groups on the poly(n-butyl acrylate) (PnBA) side chains, which are a result of ATRP, can directly react with the thiol-containing cross-linker. This direct reaction avoids the need for complex end-group transformations, streamlining the synthesis process. Secondly, the "click" thiol-bromo reaction used for cross-linking the bottlebrush polymer precursors can be completed rapidly under ambient conditions, further simplifying the operational process [178].

Moreover, the very low glass transition temperature of PnBA imparts excellent adhesiveness to the materials, enhancing their interaction with a variety of substrates. This property is particularly beneficial for applications requiring stable signal detection. For example, as a strain sensor, the PnBA bottlebrush polymer elastomers exhibit impressive sensing performance across a broad detection range of 0.5% to 100%. This combination of self-healing, adhesive properties, and wide-range sensitivity makes these materials highly versatile and promising for advanced applications in sensors and other fields requiring robust and responsive materials [178].

Further investigation into the phenomenon, the mechanical properties, and self-healing capabilities of bottlebrush polymers with dynamic and nondynamic cross-linking points were thoroughly investigated. The two types of polymers, Brush-DM (dynamic cross-linking points) and Brush-NM (non-dynamic cross-linking points), were examined to understand how the nature of cross-linking influences their performance (Fig. 5.14).

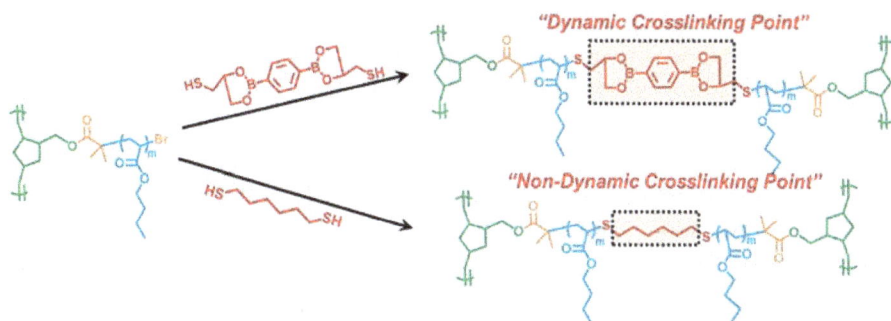

Fig. 5.14: Schematic Illustration of the synthetic route of the Brush-DM and Brush-NM featuring with the dynamic cross-linking point and nondynamic crosslinking point, respectively,, reprinted with permission from [178], copyright reserved American Chemical Society 2024.

Both Brush-DM and Brush-NM maintained a rubber-like state at room temperature, as confirmed by rheological amplitude sweeping tests. This observation is characterized by a storage modulus (G′) that significantly exceeds the loss modulus (G″) in the plateau region, indicating a well-formed elastic network. The cross-linking densities of both polymers were quantified using normalized residual thickness (NRT), showing high values: 93.6% for Brush-NM and 91.2% for Brush-DM. These high NRT values suggest that both types of polymers possess robust cross-linked networks, highlighting the effectiveness of the thiol-bromo "click" reaction used in their synthesis [178].

A key aspect of Brush-DM's performance is its dynamic cross-linking. The boronate ester bonds are not only durable but also capable of reversible rearrangement through hydrolysis and re-formation. This dynamic behavior endows Brush-DM with remarkable self-healing capabilities. In alternating step-strain sweeping tests, Brush-DM was able to recover its original rubber-like state after high strain-induced rupture. Specifically, after being subjected to a high strain of 500%, which caused a significant drop in G′, the polymer was able to restore its properties, upon reducing the strain back to 0.5%. This ability to autonomously repair and restore its mechanical properties is indicative of the effective dynamic cross-linking present in Brush-DM. In contrast, Brush-NM, with its nondynamic cross-linking, showed limited self-healing efficiency, as evidenced by the substantial decrease in modulus after cyclic testing [178].

Overall, Brush-DM stands out due to its dynamic cross-linking, which not only enhances its softness and flexibility but also imparts exceptional self-healing capabilities. The polymer's ability to recover its mechanical properties after damage and its superior softness compared to Brush-NM make it a promising candidate for applications in flexible sensors and other advanced materials. Bottlebrush polymers and their integration into self-healing polymers are evolving, as this field is gaining importance due to the discussed advantages. More examples of such developments include works by Xiong et al. [179], Shan et al. [180], Choi et al. [181], and Zha et al. [182].

Recent advancements in polymer science illustrate the broad potential of controlled polymerization techniques. For instance, self-healing coatings that exhibit both high mechanical strength and effective antifouling properties are being developed for use in challenging environments such as aerospace and marine applications [183]. These coatings leverage controlled polymerization methods to achieve precise polymer structures, which enhance their ability to repair themselves and resist environmental degradation. Similarly, in the realm of battery technology, self-healing polymer electrolytes are being designed for next-generation lithium batteries. Here, controlled polymerization enables the formation of electrolytes with specific mechanical and chemical properties that improve their longevity and safety by allowing them to self-repair after damage [184]. Additionally, super-tough and high-temperature-resistant hot-melt adhesives are being engineered through controlled polymerization techniques to meet demanding performance requirements. These adhesives benefit from the ability to fine-tune polymer properties, ensuring robust performance under extreme conditions [185]. Overall, controlled polymerization techniques are instrumental in developing polymers with targeted properties for a wide range of applications, demonstrating their critical role in advancing material performance and functionality.

5.7 Perspectives

The integration of advanced chemistry techniques into the domain of self-healing materials serves as a fundamental inception in enhancing our comprehension of their design principles and functionalities. The precise arrangement of the "building blocks" and controlled architectures play pivotal roles in achieving effective self-healing. Recent advancement in polymerization techniques (as we discussed) offers unprecedented opportunities to tailor materials with desired properties [1].

Take RAFT, for example. This technique enables the creation of custom copolymers with specific traits [186]. Copolymers synthesized through RAFT, such as furfuryl methacrylate and trifluoroethyl methacrylate, can be further refined by modifying furfuryl groups via the Diels–Alder reaction with polyhedral oligomeric silsesquioxane maleimide [187]. These modifications not only enhance hydrophobicity but also introduce thermo-reversible behavior, essential for self-healing capabilities. To date, the preparation of self-healing thermoplastics remains a relatively nascent area of research, with the majority of existing reports relying on methods that necessitate manual intervention. These methods often involve mechanisms such as thermally activated chain inter-diffusion, thermomechanical flow, and chain recombination. However, these approaches have limitations, prompting the exploration of alternative strategies [187].

In research laboratories, a novel approach has been proposed, centering on the use of living thermoplastic polymers as matrices filled with microcapsules containing monomers, serving as the healing agent provider. The use of living polymers carrying active end groups enables polymerization to resume at room temperature when fresh monomer is supplied, facilitating the covalent re-bonding of cracks through copolymerization products [188]. This approach offers the potential for both micron-scale and molecular-scale rehabilitation, thereby representing a significant advancement in the field [188]. In initial proof-of-principle experiments, living PMMA, synthesized via ATRP, was employed as the polymer matrix, with glycidyl methacrylate (GMA)-loaded microcapsules incorporated. A full recovery of impact strength at room temperature was achieved [189, 190]. However, ATRP necessitates the use of numerous catalysts, which remain in the resultant polymer and are challenging to remove [191, 192]. Moreover, exposure of ATRP products to air can lead to oxidation of the ATRP catalysts, hindering the resumption of polymerization and limiting the practical application of self-healing materials based on ATRP living polymerization.

Moreover, recent breakthroughs in RAFT polymerization have led to advancements in shape-memory and self-healing polymers. Incorporating CNTs into copolymers through RAFT, followed by metal–ligand cross-linking with Zn^{2+} ions, yields composite capable of instantaneous multiple shape-memory and precise self-healing under NIR light or heat stimulation, as we elaborated previously [193].

However, it must be noted that to classify ROMP as a typical living polymerization, several criteria must be met [195–201]. Assuming complete conversion of monomer to polymer, these criteria enable the synthesis of well-defined polymers with narrow distributions and predictable molecular weights, specified by the initial monomer-to-initiator ratio (M/I), enabling the preparation of well-defined block, graft, and other types of copolymers, end-functionalized polymers, and various other polymeric materials with complex architectures and useful functions [202].

Thus, considering the metal-mediated and equilibrium nature of most ROMP reactions, it becomes apparent that very special metathesis catalysts are needed to satisfy these requirements [199]. The catalyst should convert to growing polymer chains quantitatively and rapidly, exhibiting fast initiation kinetics. It should mediate polymerization without an appreciable amount of intramolecular or intermolecular chain transfer or premature termination. Additionally, the catalyst should react with accessible terminating agents to facilitate selective end-functionalization [199]. For practical applications, it is essential that the catalyst displays good solubility in common organic solvents or, even better, in aqueous media. Furthermore, the catalyst should show high stability toward moisture, air, and common organic functional groups to ensure robustness and reliability in various polymerization conditions [199].

Nevertheless, the study of ROMP-based self-healing systems contributes to a broader understanding of the requirements for an ideal self-healing polymer, particularly those based on a one-capsule motif. In such systems, the healing agent (a monomer) is microencapsulated as a liquid, while the polymerization initiator is embedded in the matrix as a solid [203]. Ideal characteristics of the liquid healing agent include a long shelf life, prompt deliverability, high reactivity, and low volume shrinkage upon polymerization.

However, from a scientific viewpoint, the uniqueness of autonomic self-healing systems lies in their ability to mimic natural healing processes through the integration of materials science and biochemistry [204]. These systems use sophisticated chemical reactions to repair damage without the need for external intervention. A key challenge in their development is designing chemical systems that remain inactive until triggered by specific stimuli, ensuring they are stable over time and can respond quickly when needed [205].

These self-healing systems are inspired by biological organisms, which have evolved efficient mechanisms for tissue repair. For instance, when human skin is cut, biochemical reactions stop bleeding and regenerate tissue. Autonomic self-healing materials aim to replicate this process by embedding healing agents within a matrix that releases them upon damage. These agents then react to form new bonds, effectively "healing" the material [207]. To ensure effective self-healing, several key factors must be considered. Healing agents should remain stable and dormant until activated, often encapsulated in microcapsules that rupture upon damage. Precise triggering mechanisms – such as mechanical stress, temperature changes, or chemical reactions – are essential for a rapid and efficient healing process that restores the material's original properties [208].

Moreover, the healed material must retain its mechanical integrity and functionality. This requires selecting compatible healing agents and matrix materials to ensure effective bonding. Researchers are exploring ways to improve healing efficiency by optimizing agent distribution and concentration and developing catalysts to accelerate the repair process [209]. Recent advancements include self-healing polymers, composites, and coatings. For instance, polymers with reversible covalent bonds can heal repeatedly, while composites with embedded microvascular networks transport healing agents to damaged areas, mimicking biological blood vessels. Additionally, self-healing coatings can extend the lifespan of surfaces in harsh environments, reducing maintenance costs and enhancing durability [210, 211].

For instance, consider the pioneering self-healing material crafted by researchers at the University of Illinois at Urbana-Champaign (UIUC) in 2001 [73]. By embedding microcapsules laden with liquid monomer and a polymerization catalyst within an epoxy resin matrix, the UIUC team achieved a self-healing mechanism reminiscent of biological wound healing [73]. Upon the formation of microcracks, the ruptured microcapsules release their contents, triggering a polymerization reaction that fills and bonds the cracks, restoring structural integrity.

The chemistry circumscribing this self-healing process draws from olefin metathesis, a versatile reaction honored with the 2005 Nobel Prize in Chemistry. Specifically, ring-opening metathesis polymerization catalyzed by Grubbs-type catalysts swiftly converts strained cyclic alkenes, like dicyclopentadiene, into cross-linked polymeric networks, offering an elegant solution for efficient self-healing, while maintaining compatibility with epoxy resin matrices [212]. Remarkably, the UIUC team's material showcased exceptional resilience, with self-healing efficiency tests revealing that over 75% of the material's mechanical strength could be restored. Moreover, innovative tactics, such as integrating microvascular networks, enabled repeated healing at the same location, overcoming the challenge of depleted chemical potential upon healing agent polymerization [73].

Additionally, spontaneous, nonmetal-mediated chemistries have shown promise in facilitating self-healing processes. For instance, nucleophilic amines can react with electrophilic epoxides to form cross-linked polymeric networks, offering an alternative approach to autonomous healing. By encapsulating these reactive chemical agents within hollow fibers dispersed throughout the material, researchers have achieved rapid and efficient healing responses upon microcrack formation [32].

Nevertheless, it is true that CLP techniques remain one of the most fascinating approaches to developing advanced polymers. These techniques allow for precise control over polymer architecture and molecular weight, which is crucial for creating polymers with specific functionalities. In addition to providing access to materials with reprocessability, RAFT-mediated 3D printing has enabled the manufacturing of materials with controlled nanoscale structural features via the polymerization-induced microphase separation (PIMS) process [150]. This process, initially reported by Hillmyer's group, involves chain-extending a macromolecular chain-transfer agent (macro-CTA) that is initially soluble in a polymerization mixture containing a multi-functional monomer. As the chain extends, the macro-CTA becomes incompatible with the polymerizing medium, leading to microphase separation and thus the formation of nanostructures [213].

The chain extension via RAFT polymerization allows for linear increases in molar mass and dictates the resultant morphology, enabling control over the nanostructuration. For example, the Boyer group recently employed this technique in 3D printing, which allows for the modulation of the morphology of nanostructured 3D printed materials by adjusting the macro-CTA chain length, architecture, and loading [214]. This capability was previously impossible with conventional 3D printing methods. Their studies have shown that 3D-printed materials undergo morphological modifications as the chain length of macro-CTAs increases while maintaining the loading. The materials transform from those with isolated globular domains to those with elongated domains and eventually, bicontinuous morphologies.

In the context of self-healing polymers, these advancements are particularly significant. The precise control over polymer morphology, enabled by RAFT polymerization and 3D printing techniques, allows for the design of materials that can respond to damage with a high degree of specificity and efficiency.

Moreover, the integration of self-healing mechanisms into 3D-printed structures enables the creation of complex geometries that can maintain their integrity over time, even under mechanical stress. This is particularly relevant for applications in flexible electronics where materials are subjected to repeated bending and stretching. The ability to heal microcracks and other forms of damage in these materials ensures their continued functionality and durability. Combining CLP with 3D printing is an effective way to create self-healing polymers with well-defined nanostructures and specific properties.

5.8 Conclusion

CLPs offer significant advantages in the development of self-healing polymers, primarily through their ability to control molecular weight and incorporate diverse functional groups.

Firstly, CLPs provide precise control over the molecular weight of polymer chains, which is essential for tailoring the properties of the final material. Adjusting the molecular weight directly affects characteristics such as chain length, flexibility, strength, and durability [70]. This fine-tuning enables scientists to design polymers with specific attributes, enhancing self-healing capabilities, mechanical properties, or thermal stability (Fig. 5.15) [70].

Secondly, CLPs facilitate the incorporation of various functional groups along the polymer backbone, allowing for customized material properties. For instance, adding groups that form dynamic covalent bonds can improve the polymer's self-healing ability, enabling autonomous repair, upon damage. Other functional groups can elicit specific responses to external stimuli, such as heat, light, or pH changes, thereby expanding the range of applications for these materials (Fig. 5.15) [215]. The precision and versatility of CLPs make them powerful tools in developing advanced self-healing polymers, creating robust and adaptive materials suitable for a variety of environmental conditions.

Self-healing polymers have significant potential in extreme environments, offering smart material solutions in design and engineering. They reduce the need for human intervention in repairing critical components, especially in high-risk situations. Additionally, their use can lower repair costs for inaccessible or remote components, extending service life and maintaining operational efficiency [215]. While current extrinsic self-healing methods, which use microcapsule-loaded healing agents, show promise, they often suffer from irreversibility and are limited to a single healing cycle [1]. Conversely, extrinsic microvascular networks allow for multiple healing events but require complex and costly fabrication. In contrast, intrinsic self-healing mechanisms, based on the macromolecular design of polymer structures, offer high efficiencies and potentially infinite healing cycles through dynamic covalent bonds or noncovalent interactions, providing versatile opportunities for intelligent and biomimetic design [1].

As we reviewed, recent progress in synthetic polymer chemistry, especially the development of RDRP methods, has greatly improved the ability to precisely control polymer structures. Despite these advancements, there is still a significant challenge in applying these techniques to create polymers that can operate and self-heal in extreme environments [216].

We believe that each extreme environment poses unique challenges that demand tailored solutions within polymer design to enable efficient self-healing. For instance, low-temperature environments present kinetic hurdles for reaction rates, while environments like the deep ocean necessitate self-healing mechanisms resistant to water molecules, high salinity, and high pressure [217]. With limited exploration in self-healing materials for extreme environments, numerous pathways remain uncharted, offering fertile ground for the development of novel structures and chemistries to extend healing mechanisms into extreme conditions [218].

Furthermore, the physical, thermal, and viscoelastic properties of polymers must be meticulously engineered to suit the demands of each extreme environment [219]. For instance, polymers intended for high-intensity UV light environments must exhibit resilience against photo-induced chain scission. However, the distinct conditions of each extreme environment also present opportunities for innovating new self-healing mechanisms, leveraging thermal energy, UV light, or magnetic fields [219].

To reiterate, several methods have been developed for the synthesis of functionalized polymers. These include ionic and free radical polymerization of vinyl monomers, group transfer polymerization (GTP), and, more recently, ROMP. ROMP is particularly noteworthy due to its reliability, its ability to produce linear materials, and its compatibility with forming various copolymers with controlled architectures [196]. Polymers made using ROMP can include different chelating groups, such as terpyridine moieties, which act as anchor points for binding metal complexes to the polymer chain [197].

However, maintaining the "living" nature of ROMP, which allows for continuous polymer growth without termination, remains a challenge. It is important to develop systems that enable better control over this aspect [220]. Additionally, efforts should be directed toward understanding how these systems can be integrated into high-performance self-healing polymers.

Recent developments in polymer chemistry have led to interesting innovations. For example, Jeong and colleagues recently created a polymer based on terpyridine and norbornene using ROMP. They demonstrated that this polymer could function as a regenerative and repairable photocatalyst. This polymer, activated by visible light, can catalyze the oxidation of benzylamine into imine. The polymer includes a metal-complex component that binds to a ruthenium(II) complex, allowing it to act as a reversible photocatalyst [221].

Despite these advancements, terpyridine-based norbornene polymers have not yet been studied for their potential in self-healing applications, which represents an intriguing area for future exploration [221, 222]. The ability of these polymers to incorporate metal-binding chelating groups like terpyridine suggests they could be tailored for responsive or dynamic behaviors under various stimuli, such as light, heat, or chemical environments. This potential for responsiveness is a promising avenue for developing materials with self-healing properties.

Controlled polymerization techniques, such as RAFT and ATRP, offer notable advantages from an industrial perspective. These techniques have been extensively discussed by researchers like Dr. Destarac [223] and Dr. Anastasaki [133] and colleagues, highlighting their potential for large-scale development.

A major challenge for both RAFT and ATRP polymerization methods is the balance between cost and performance [133, 227]. While these techniques can produce polymers with superior properties and functionalities, their higher production costs mean that they are often reserved for applications where the advanced qualities of the polymers justify the expense. Therefore, it is crucial to ensure that materials developed using these techniques offer sufficient longevity and durability to make the higher costs worthwhile. In other words, the benefits of the advanced properties should be significant enough to offset the initial investment in these high-cost polymers.

Advanced self-healing polymers can undergo multiple fracture-healing cycles with minimal reduction in efficiency, making them ideal for applications requiring long-lasting durability and self-repair [228]. This capability is particularly beneficial for critical systems that experience frequent wear and tear, as it can significantly extend the lifespan and reliability of components [228].

However, a key limitation of some self-healing polymers is that their healing process often necessitates heating. This requirement can pose challenges in practical applications, especially in environments where temperature control is difficult. Future research should focus on developing self-healing polymers that operate effectively without external heat or on creating systems that can efficiently manage the required heating during the healing process.

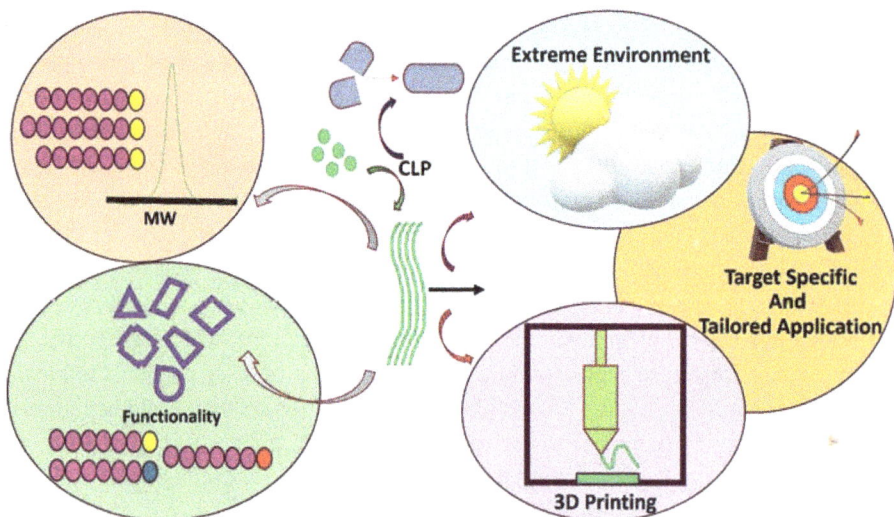

Fig. 5.15: Summarization and illustration of the advantages of the CLP synthetic route and its future outlook.

Despite this constraint, the development of such materials highlights the ongoing innovation in the field of self-healing polymers and broadens the scope of potential solutions beyond traditional CLP-based approaches. Thus, for self-healing technologies to become commercially viable, especially with the CLP synthetic strategy, it is essential to move from academic research to industrial use. This involves creating prototypes that are ready for practical application, such as self-healing materials for aircraft or marine structures. The development process should also focus on selecting appropriate chemical methods and production techniques that allow for large-scale manufacturing, like producing corrosion-resistant coatings for ships.

Looking ahead, self-healing polymers have potential applications in various areas, including wireless sensors, self-powered sensors, energy generators, and soft robotics. These materials could be used in challenging environments such as oceans, oil fields, and space. Self-healing polymer-based composites can offer lightweight and cost-effective solutions for these applications. By combining different approaches in polymer composites and nanocomposites, the effectiveness of self-healing can be improved, opening up new possibilities for energy generation and sensor technology in extreme conditions. Addressing the technical challenges of self-healing in such environments will help advance polymer science and move these technologies closer to practical use.

References

[1]   Yang, Y. & Urban, M. W. (2013). *Chemical Society Reviews, 42*, 7446–7467.
[2]   Venkateswaran, M. R., Khosravi, A., Zarepour, A., Iravani, S. & Zarrabi, A. (2024). *Environmental Science: Nanotechnology.*
[3]   Wu, D. Y., Meure, S. & Solomon, D. (2008). *Progress in Polymer Science, 33*, 479–522.
[4]   Li, B., Cao, P. F., Saito, T. & Sokolov, A. P. (2022). *Chemical Reviews, 123*, 701–735.
[5]   Trask, R. S., Williams, H. R. & Bond, I. P. (2007). *Bioinspiration & Biomimetics, 2*, P1.
[6]   Cremaldi, J. C. & Bhushan, B. (2018). *Beilstein Journal of Nanotechnology, 9*, 907–935.
[7]   Khaneghahi, M. H., Kamireddi, D., Rahmaninezhad, S. A., Sadighi, A., Schauer, C. L., Sales, C. M. & Farnam, Y. A. (2023). *Construction and Building Materials, 408*, 133765.
[8]   Yang, Y., Davydovich, D., Hornat, C. C., Liu, X. & W, M. (2018). Urban, Chemistry, *4*, 1928–1936.
[9]   D'elia, E., Eslava, S., Miranda, M., Georgiou, T. K. & Saiz, E. (2016). *Scientific Reports, 6*, 25059.
[10]   Zhang, H., Zhang, X., Bao, C., Li, X., Duan, F., Friedrich, K. & Yang, J. (2019). *Chemicals Materials, 31*, 2611–2618.
[11]   Speck, O. & Speck, T. (2019). *Biomimetics, 4*, 26.
[12]   Li, C. H. & Zuo, J. L. (2020). *Advances in Materials, 32*, 1903762.
[13]   Sirajuddin, N. A. & Jamil, M. M. (2015). *Sains Malaysiana, 44*, 811–818.
[14]   Imato, K., Takahara, A. & Otsuka, H. (2015). *Macromolecules, 48*, 5632–5639.
[15]   Yang, Y., Ding, X. & Urban, M. W. (2015). *Prog. Polymer Science, 49*, 34–59.
[16]   Grabowski, B. & Tasan, C. C. (2016). *Self-Healing Materials*, 387–407.
[17]   Zhang, S., van Dijk, N. & van der Zwaag, S. (2020). *Acta Metallurgica Sinica (Engl. Lett.), 33*, 1167–1179.
[18]   Ferguson, J. B., Schultz, B. F. & Rohatgi, P. K. (2014). *JOM, 66*, 866–871.
[19]   van Dijk, N. & van der Zwaag, S. (2018). *Advanced Materials Interfaces, 5*, 1800226.

[20]   Wu, Y., Huang, L., Huang, X., Guo, X., Liu, D., Zheng, D. & Chen, J. (2017). *Energy & Environmental Science*, *10*, 1854–1861.

[21]   Wu, Y., Huang, X., Huang, L., Guo, X., Ren, R., Liu, D. & Chen, J. (2018). *ACS Applied Energy Materials*, *1*, 1395–1399.

[22]   Huang, C., Wang, X., Cao, Q., Zhang, D. & Jiang, J. Z. (2021). *ACS Applied Energy Materials*, *4*, 12224–12231.

[23]   Zhang, H., Chen, P., Xia, H., Xu, G., Wang, Y., Zhang, T. & Sun, Z. (2022). *Energy & Environmental Science*, *15*, 5240–5250.

[24]   Greil, P. (2020). *Advanced Engineering Materials*, *22*, 1901121.

[25]   Hager, M. D., Greil, P., Leyens, C., van Der Zwaag, S. & Schubert, U. S. (2010). *Advances in Materials*, *22*, 5424–5430.

[26]   Nakao, W., Takahashi, K. & Ando, K. (2009). *Self-Healing Materials*, 183–217.

[27]   Madhan, M. & Prabhakaran, G. *Int. Conf. Intell. Robotics, Automation, and Manufacturing*. Springer Berlin Heidelberg: Berlin, Heidelberg, 2012, pp. 466–474.

[28]   Ozaki, S., Osada, T. & Nakao, W. (2016). *International Journal of Solids and Structures*, *100*, 307–318.

[29]   Herbst, F., Döhler, D., Michael, P. & Binder, W. H. (2013). *Macromolecular Rapid Communications*, *34*, 203–220.

[30]   Reddy, K. R., El-Zein, A., Airey, D. W., Alonso-Marroquin, F., Schubel, P. & Manalo, A. (2020). *Nano-Struct. Nano-Objects*, *23*, 100500.

[31]   Luo, J., Wang, T., Sim, C. & Li, Y. (2022). *Polymers*, *14*(14), 2808.

[32]   Dai, L. H., Wu, C., An, F. J. & Liao, S. S. (2018). *Advance Material Science and Engineering*, *2018*(1), 1264276.

[33]   Samadzadeh, M., Boura, S. H., Peikari, M., Kasiriha, S. M. & Ashrafi, A. (2010). *Progress in Organic Coatings.*, *68*(3), 159–164.

[34]   Zhang, F., Ju, P., Pan, M., Zhang, D., Huang, Y., Li, G. & Li, X. (2018). *Corrosion Science*, *144*, 74–88.

[35]   Rule, J. D., Sottos, N. R. & White, S. R. (2007). *Polymer*, *48*(12), 3520–3529.

[36]   Ekeocha, J., Ellingford, C., Pan, M., Wemyss, A. M., Bowen, C. & Wan, C. (2021). *Advances in Materials*, *33*(33), 2008052.

[37]   Hillewaere, X. K. & Du Prez, F. E. (2015). *Progress in Polymer Science*, *49*, 121–153.

[38]   Wang, Z., Lu, X., Sun, S., Yu, C. & Xia, H. (2019). *Journal of Materials Chemical B*, *7*(32), 4876–4926.

[39]   Pratama, P. A., Sharifi, M., Peterson, A. M. & Palmese, G. R. (2013). *ACS Applied Materials & Interfaces*, *5*(23), 12425–12431.

[40]   Ling, L., Li, J., Zhang, G., Sun, R. & Wong, C. P. (2018). *Macromolecular Research*, *26*(4), 365–373.

[41]   Cao, J., Lu, C., Zhuang, J., Liu, M., Zhang, X., Yu, Y. & Tao, Q. (2017). *Angewandte Chemie*, *129*(30), 8921–8926.

[42]   Agrawal, N. & Arora, B. (2022). *Mini-Reviews in Organic Chemistry*, *19*(4), 496–512.

[43]   Malekkhouyan, R., Neisiany, R. E., Khorasani, S. N., Das, O., Berto, F. & Ramakrishna, S. (2021). *Journal of Applied Polymer Science*, *138*(10), 49964.

[44]   Parameswaran, B. & Singha, N. K. *Toughened Composites*. CRC Press. 2022. pp. 203–216.

[45]   Liu, Z., Zhong, Y., Li, S., Yu, S., Zhong, J., Yang, Y. & Shen, L. (2024). *Macromolecular Chemistry and Physics*, 2400079.

[46]   AbdolahZadeh, M., Van Der Zwaag, S. & Garcia, S. J. (2016). *Self-Healing Materials*, 185–218.

[47]   Zhang, S., van Dijk, N. & van der Zwaag, S. (2020). *Acta Metallurgica Sinica (Engl. Lett.)*, *33*, 1167–1179.

[48]   Binder, W. H. (ed.) *Self-Healing Polymers: From Principles to Applications*. John Wiley & Sons, 2013.

[49]   Ghosh, S. K. (ed.) *Self-Healing Materials: Fundamentals, Design Strategies, and Applications*. Wiley-VCH, Vol. 18, 2009.

[50]   Hager, M. D., Greil, P., Leyens, C., Van Der Zwaag, S. & Schubert, U. S. (2010). *Advances in Materials*, *22*(47), 5424–5430.

[51]   Roy, N., Bruchmann, B. & Lehn, J. M. (2015). *Chemical Society Reviews*, *44*(11), 3786–3807.

[52] Guimard, N. K., Oehlenschlaeger, K. K., Zhou, J., Hilf, S., Schmidt, F. G. & Barner-Kowollik, C. (2012). *Macromolecular Chemistry and Physics*, *213*(2), 131–143.

[53] Song, T., Jiang, B., Li, Y., Ji, Z., Zhou, H., Jiang, D. & Colorado, H. (2021). *ES Materials & Manufacturing*, *14*, 1–19.

[54] Xiao, D. S., Yuan, Y. C., Rong, M. Z. & Zhang, M. Q. (2009). *Advanced Functional Materials*, *19*(14), 2289–2296.

[55] Huynh, T. P., Sonar, P. & Haick, H. (2017). *Advances in Materials*, *29*(19), 1604973.

[56] Hia, I. L., Vahedi, V. & Pasbakhsh, P. (2016). *Polymer Reviews*, 56(2), 225–261.

[57] Scheiner, M., Dickens, T. J. & Okoli, O. (2016). *Polymer*, *83*, 260–282.

[58] Cioffi, M. O. H., Bomfim, A. S., Ambrogi, V. & Advani, S. G. (2022). *Polymer Compos*, *43*(11), 7643–7668.

[59] Dahlke, J., Zechel, S., Hager, M. D. & Schubert, U. S. (2018). *Advanced Materials Interfaces*, *5*(17), 1800051.

[60] Xie, J., Yu, P., Wang, Z. & Li, J. (2022). *Biomacromolecules*, *23*(3), 641–660.

[61] Brown, E. N., White, S. R. & Sottos, N. R. (2004). *Journal of Materials Science*, *39*, 1703–1710.

[62] Keller, M. W. & Sottos, N. R. (2006). *Experimental Mechanics*, *46*, 725–733.

[63] Jones, A. S., Rule, J. D., Moore, J. S., White, S. R. & Sottos, N. R. (2006). *Chemical Materials*, *18*(5), 1312–1317.

[64] Toohey, K. S., Sottos, N. R., Lewis, J. A., Moore, J. S. & White, S. R. (2007). *Nature Materials*, *6*(8), 581–585.

[65] Cho, S. H., Andersson, H. M., White, S. R., Sottos, N. R. & Braun, P. V. (2006). *Advances in Materials*, *18*(8), 997–1000.

[66] Zhong, N. & Post, W. (2015). *Composites: Part A Applied Science and Manufacturing*, *69*, 226–239.

[67] Hayes, S. A., Zhang, W., Branthwaite, M. & Jones, F. R. (2007). *Journal of the Royal Society Interface*, *4*(13), 381–387.

[68] Scheiner, M., Dickens, T. J. & Okoli, O. (2016). *Polymer*, *83*, 260–282.

[69] Williams, K. A., Dreyer, D. R. & Bielawski, C. W. (2008). *MRS Bullettin*, *33*(8), 759–765.

[70] Corrigan, N., Jung, K., Moad, G., Hawker, C. J., Matyjaszewski, K. & Boyer, C. (2020). *Progress in Polymer Science*, *111*, 101311.

[71] Utrera-Barrios, S., Verdejo, R., López-Manchado, M. A. & Santana, M. H. (2020). *Materials Horizons*, *7*(11), 2882–2902.

[72] Dry, C. (1992). *International Journal of Modern Physics B*, *6*.

[73] White, S. R., Sottos, N. R., Geubelle, P. H., Moore, J. S., Kessler, M. R., Sriram, S. R. & Viswanathan, S. (2001). *Nature*, *409*(6822), 794–797.

[74] van der Zwaag, S. & Brinkman, E. (ed.) *Self-Healing Materials*. Springer Netherlands, 2008, pp. 63–76.

[75] Cordier, P., Tournilhac, F., Soulié-Ziakovic, C. & Leibler, L. (2008). *Nature*, *451*, 977–980.

[76] Toohey, K. S., Sottos, N. R., Lewis, J. A., Moore, J. S. & White, S. R. (2007). *Nature Materials*, *6*, 581–585.

[77] Roy, S. G. & De, P. (2020). *Reversible Deactivation Radical Polymerization: Synthesis and Applications of Functional Polymers*. 221.

[78] Wang, H. S., Parkatzidis, K., Junkers, T., Truong, N. P. & Anastasaki, A. (2024). *Chemicals*, *10*, 388–401.

[79] Krys, P., Ribelli, T. G., Matyjaszewski, K. & Gennaro, A. (2016). *Macromolecules*, *49*, 2467–2476.

[80] Krys, P. & Matyjaszewski, K. (2017). *European Polymer Journal*, *89*, 482–523.

[81] Matyjaszewski, K., Patten, T. E. & Xia, J. (1997). *Journal of the American Chemical Society*, *119*, 674–680.

[82] Pearson, S., Thomas, C. S., Guerrero-Santos, R. & d'Agosto, F. (2017). *Polymer Chemistry*, *8*, 4916–4946.

[83] Yildirim, I., Weber, C. & Schubert, U. S. (2018). *Progress in Polymer Science*, *76*, 111–150.

[84] Wang, J. L., Grimaud, T. & Matyjaszewski, K. (1997). *Macromolecules*, *30*, 6507–6512.

[85] Zhang, H., Klumperman, B., Ming, W., Fischer, H. & van der Linde, R. (2001). *Macromolecules*, *34*, 6169–6173.

[86] Li, Y. M., Zhang, Z. P., Rong, M. Z. & Zhang, M. Q. (2022). *Nature Communications*, *13*, 2633.
[87] Tan, Y. J., Wu, J., Li, H. & Tee, B. C. (2018). *ACS Applied Materials and Interfaces*, *10*, 15331–15345.
[88] Bekas, D. G., Tsirka, K., Baltzis, D. & Paipetis, A. S. (2016). *Composites Part B: Engineering*, *87*, 92–119.
[89] Lee, M. W., An, S., Yoon, S. S. & Yarin, A. L. (2018). *Advances in Colloid Interface Science*, *252*, 21–37.
[90] Derboven, P. *Doctoral dissertation*. Ghent University, 2016.
[91] Ejeromedoghene, O., Abesa, S., Akor, E. & Omoniyi, A. O. (2023). *Materials Today Communications*, *35*, 106063.
[92] De, P., Gondi, S. R., Roy, D. & Sumerlin, B. S. (2009). *Macromolecules*, *42*, 5614–5621.
[93] Ihsan, A. B., Sun, T. L., Kurokawa, T., Karobi, S. N., Nakajima, T., Nonoyama, T. & Gong, J. P. (2016). *Macromolecules*, *49*, 4245–4252.
[94] Pramanik, N. B., Nando, G. B. & Singha, N. K. (2015). *Polymer*, *69*, 349–356.
[95] Yao, L., Yuan, Y. C., Rong, M. Z. & Zhang, M. Q. (2011). *Polymer*, *52*, 3137–3145.
[96] Bobrin, V. A., Zhang, J., Corrigan, N. & Boyer, C. (2023). *Advanced Materials Technologies*, *8*, 2201054.
[97] Aydogan, C., Yilmaz, G., Shegiwal, A., Haddleton, D. M. & Yagci, Y. (2022). *Angewandte Chemie*, *134*, e202117377.
[98] Jung, K., Xu, J., Zetterlund, P. B. & Boyer, C. (2015). *ACS Macro Letters*, *4*, 1139–1143.
[99] Wang, L., Xu, Y., Zuo, Q., Dai, H., Huang, L., Zhang, M. & Zhou, Y. (2022). *Nature Communications*, *13*, 3621.
[100] Zhang, Z., Corrigan, N., Jin, A. & Boyer, C. (2019). *Angewandte Chemie International Edition*, *58*, 17954–17963.
[101] Zhang, Z., Corrigan, N. & Boyer, C. (2021). *Macromolecules*, 54, 1170–1182.
[102] Shi, X., Zhang, J., Corrigan, N. & Boyer, C. (2021). *Materials Chemistry Frontiers*, *5*, 2271–2282.
[103] Zhao, B., Li, J., Pan, X., Zhang, Z., Jin, G. & Zhu, J. (2021). *ACS Macro Letters*, *10*, 1315–1320.
[104] Qiao, L., Zhou, M., Shi, G., Cui, Z., Zhang, X., Fu, P. & Pang, X. (2022). *Journal of the American Chemical Society*, *144*, 9817–9826.
[105] Mohapatra, H., Kleiman, M. & Esser-Kahn, A. P. (2017). *Nature Chemistry*, *9*, 135–139.
[106] Zaborniak, I., Chmielarz, P., Wolski, K., Grześ, G., Wang, Z., Górska, A. & Matyjaszewski, K. (2022). *European Polymer Journal*, *164*, 110972.
[107] Zhou, M., Zhang, Y., Shi, G., He, Y., Cui, Z., Zhang, X. & Pang, X. (2022). *ACS Macro Letters*, *12*, 26–32.
[108] Feng, H., Shao, X. & Wang, Z. (2024). *ChemPlusChem*, e202400287.
[109] Ren, Z., Ding, C., Ding, R., Wang, J., Li, Z., Tan, R. & Zhang, Z. (2023). *ACS Macro Letters*, *12*, 1159–1165.
[110] Magenau, A. J., Strandwitz, N. C., Gennaro, A. & Matyjaszewski, K. (2011). *Science*, *332*, 81–84.
[111] Strover, L. T., Cantalice, A., Lam, J. Y., Postma, A., Hutt, O. E., Horne, M. D. & Moad, G. (2019). *ACS Macro Letters*, *8*, 1316–1322.
[112] Lorandi, F., Fantin, M., Shanmugam, S., Wang, Y., Isse, A. A., Gennaro, A. & Matyjaszewski, K. (2019). *Macromolecules*, *52*, 1479–1488.
[113] Bray, C., Li, G., Postma, A., Strover, L. T., Wang, J. & Moad, G. (2020). *Australian Journal of Chemistry*, *74*, 56–64.
[114] Zhang, L., Laborda, E., Darwish, N., Noble, B. B., Tyrell, J. H., Pluczyk, S. & Ciampi, S. (2018). *Journal of the American Chemical Society*, *140*, 766–774.
[115] Zaborniak, I., Chmielarz, P., Wolski, K., Grześ, G., Wang, Z., Górska, A. & Matyjaszewski, K. (2022). *European Polymer Journal*, *164*, 110972.
[116] Barthel, M. J., Rudolph, T., Teichler, A., Paulus, R. M., Vitz, J., Hoeppener, S. & Schubert, U. S. (2013). *Advanced Functional Materials*, *23*, 4921–4932.
[117] Zheng, P. & McCarthy, T. J. (2012). *Journal of the American Chemical Society*, *134*, 2024–2027.
[118] Barthel, M. J., Rudolph, T., Crotty, S., Schacher, F. H. & Schubert, U. S. (2012). *Journal of Polymer Science Part A: Polymer Chemistry*, *50*, 4958–4965.
[119] Xie, J., Gao, L., Yan, L., Wang, R., Yang, Y., Hu, M. & Wang, J. (2015). *Polymer Chemistry*, *6*, 5753–5762.
[120] Xiao, D. S., Yuan, Y. C., Rong, M. Z. & Zhang, M. Q. (2009). *Polymer*, *50*(13), 2967–2975.

[121] Hondred, P. R., Autori, C. & Kessler, M. R. (2014). *Macromolecular Materials and Engineering, 299*(9), 1062–1069.

[122] Guo, W., Jia, Y., Tian, K., Xu, Z., Jiao, J., Li, R. & Wang, H. (2016). *ACS Applied Materials and Interfaces, 8*(32), 21046–21054.

[123] Döhler, D., Zare, P. & Binder, W. H. (2014). *Polymer Chemistry, 5*(3), 992–1000.

[124] Hondred, P. R. *Doctoral dissertation. Iowa State University*, 2013.

[125] Mauldin, T. C., Leonard, J., Earl, K., Lee, J. K. & Kessler, M. R. (2012). *ACS Applied Materials and Interfaces, 4*(3), 1831–1837.

[126] McIlroy, D. A., Blaiszik, B. J., Caruso, M. M., White, S. R., Moore, J. S. & Sottos, N. R. (2010). *Macromolecules*, 43(4), 1855–1859.

[127] Billiet, S., Van Camp, W., Hillewaere, X. K., Rahier, H. & Du Prez, F. E. (2012). *Polymer, 53*(12), 2320–2326.

[128] Huang, L., Yang, S., Chen, J., Tian, J., Huang, Q., Huang, H. & Wei, Y. (2019). *Materials Science and Engineering: C, 94*, 270–278.

[129] Jiang, R., Liu, M., Huang, H., Mao, L., Huang, H., Wen, Y. & Wei, Y. (2018). *Journal of Colloid and Interface Science, 519*, 137–144.

[130] Tian, J., Jiang, R., Gao, P., Xu, D., Mao, L., Zeng, G. & Wei, Y. (2017). *Materials Science and Engineering: C, 79*, 563–569.

[131] Long, Z., Liu, M., Wang, K., Deng, F., Xu, D., Liu, L. & Wei, Y. (2016). *Materials Science and Engineering: C, 66*, 215–220.

[132] Herbst, F., Seiffert, S. & Binder, W. H. (2012). *Polymer Chemistry, 3*(11), 3084–3092.

[133] Truong, N. P., Jones, G. R., Bradford, K. G., Konkolewicz, D. & Anastasaki, A. (2021). *Nature Reviews Chemistry, 5*(12), 859–869.

[134] Chiefari, J., Chong, Y. K., Ercole, F., Krstina, J., Jeffery, J., Le, T. P. & Thang, S. H. (1998). *Macromolecules, 31*(16), 5559.

[135] Nothling, M. D., Fu, Q., Reyhani, A., Allison-Logan, S., Jung, K., Zhu, J. & Qiao, G. G. (2020). *Advanced Science, 7*(20), 2001656.

[136] Lee, Y., Boyer, C. & Kwon, M. S. (2023). *Chemical Society Reviews*, 52(9), 3035–3097.

[137] Moad, G. (2017). *Polymer Chemistry, 8*(1), 177–219.

[138] Boyer, C., Bulmus, V., Davis, T. P., Ladmiral, V., Liu, J. & Perrier, S. (2009). *Chemical Reviews, 109*(11), 5402–5436.

[139] Moad, G. & Rizzardo, E. (eds.). *RAFT Polymerization: Methods, Synthesis, and Applications*. John Wiley & Sons, 2021.

[140] Harrisson, S., Liu, X., Ollagnier, J. N., Coutelier, O., Marty, J. D. & Destarac, M. (2014). *Polymers, 6*(5), 1437–1488.

[141] Ponnupandian, S., Mondal, P., Becker, T., Hoogenboom, R., Lowe, A. B. & Singha, N. K. (2021). *Polymer Chemistry, 12*(6), 876–884.

[142] Gong, X., Cheng, Z., Gao, S., Zhang, D., Ma, Y., Wang, J. & Chu, F. (2020). *Carbohydrate Polymers, 250*, 116846.

[143] Li, H., Zhou, J. & Zhao, J. (2023). *Applied Surface Science, 614*, 156180.

[144] Wang, F., Wang, W., Zhang, C., Tang, J., Zeng, X. & Wan, X. (2021). *Composites Part B: Engineering, 219*, 108927.

[145] Zhao, J., Diaz-Dussan, D., Wu, M., Peng, Y. Y., Wang, J., Zeng, H. & Narain, R. (2020). *Biomacromolecules, 22*(2), 800–810.

[146] Capets, J. A., Yost, S. F., Vogt, B. D. & Pester, C. W. (2024). *Advanced Functional Materials*, 2406277.

[147] Rigby, A. D., Alipio, A. R., Chiaradia, V. & Arno, M. C. (2023). *Biomacromolecules, 24*(7), 3370–3379.

[148] Li, H., Zhou, J. & Zhao, J. (2023). *Applied Surface Science, 614*, 156180.

[149] Zha, H., Cheng, L., Wang, Z., Liu, C. & Hong, C. (2024). *Science China Chemicals*, 1–9.

[150] Bagheri, A. (2023). *Macromolecules, 56*(5), 1778–1797.

[151] Zhang, Z., Corrigan, N. & Boyer, C. (2022). *Angewandte Chemie, 134*(11), e202114111.

[152] Zhao, B., Li, J., Xiu, Y., Pan, X., Zhang, Z. & Zhu, J. (2022). *Macromolecules, 55*(5), 1620–1628.

[153] Le, D., Morandi, G., Legoupy, S., Pascual, S., Montembault, V. & Fontaine, L. (2013). *European Polymer Journal, 49*(5), 972–983.

[154] Arslan, M. & Tasdelen, M. A. (2019). *Chemistry Africa, 2*(2), 195–214.

[155] Matyjaszewski, K. (2012). *Macromolecules, 45*(10), 4015–4039.

[156] Xia, J., Gaynor, S. G. & Matyjaszewski, K. (1998). *Macromolecules, 31*(17), 5958–5959.

[157] Fan, D., Wang, W., Chen, H., Bai, L., Yang, H., Wei, D. & Niu, N. Y. (2019). *New Journal of Chemistry, 43*(7), 3099–3110.

[158] Bai, L., Jiang, X., Sun, Z., Pei, Z., Ma, A., Wang, W. & Wei, D. (2019). *Cellulose, 26*, 5305–5319.

[159] Ma, A., Zhang, J., Wang, N., Bai, L., Chen, H., Wang, W. & Wei, D. (2018). *Industrial & Engineering Chemistry Research, 57*(51), 17417–17429.

[160] Ma, A., Jiang, C., Li, M., Cao, L., Deng, Z., Bai, L. & Wei, D. (2020). *Reactive and Functional Polymers, 150*, 104547.

[161] Jiang, X., Xi, M., Bai, L., Wang, W., Yang, L., Chen, H. & Wei, D. (2020). *Materials Science and Engineering: C, 109*, 110553.

[162] Zhu, D. Y., Rong, M. Z. & Zhang, M. Q. (2015). *Progress in Polymer Science, 49*, 175–220.

[163] Wang, S. & Urban, M. W. (2020). *Nature Reviews Materials, 5*(8), 562–583.

[164] Jiang, X., Xi, M., Bai, L., Wang, W., Yang, L., Chen, H. & Wei, D. (2020). *Materials Science and Engineering: C, 109*, 110553.

[165] Kord Forooshani, P. & Lee, B. P. (2017). *Journal of Polymer Science Part A: Polymer Chemistry, 55*(1), 9–33.

[166] Zhao, Y., Yin, R., Wu, H., Wang, Z., Zhai, Y., Kim, K. & Bockstaller, M. R. (2023). *ACS Macro Letters, 12*(4), 475–480.

[167] Urban, M. W., Davydovich, D., Yang, Y., Demir, T., Zhang, Y. & Casabianca, L. (2018). *Science, 362*(6411), 220–225.

[168] Urban, M. W. *U.S. Patent No. 11,312,807.* U.S. Patent and Trademark Office: Washington, DC, 2022.

[169] Li, Z., Tang, M., Liang, S., Zhang, M., Biesold, G. M., He, Y. & Lin, Z. (2021). *Progress in Polymer Science, 116*, 101387.

[170] Paturej, J., Sheiko, S. S., Panyukov, S. & Rubinstein, M. (2016). *Science Advances, 2*(11), e1601478.

[171] Abbasi, M., Faust, L. & Wilhelm, M. (2019). *Advances in Materials, 31*(26), 1806484.

[172] Verduzco, R., Li, X., Pesek, S. L. & Stein, G. E. (2015). *Chemical Society Reviews, 44*(8), 2405–2420.

[173] Dalsin, S. J., Rions-Maehren, T. G., Beam, M. D., Bates, F. S., Hillmyer, M. A. & Matsen, M. W. (2015). *ACS Nanotechnology, 9*(12), 12233–12245.

[174] Xiong, H., Yue, T., Wu, Q., Zhang, L., Xie, Z., Liu, J. & Wu, J. (2023). *Materials Horizons, 10*(6), 2128–2138.

[175] Xiao, L., Chen, Y. & Zhang, K. (2016). *Macromolecules, 49*(12), 4452–4461.

[176] Lin, T. P., Chang, A. B., Luo, S. X. L., Chen, H. Y., Lee, B. & Grubbs, R. H. (2017). *ACS Nanotechnology, 11*(11), 11632–11641.

[177] Hourston, D. J. & Schäfer, F. U. (1996). *Polymers for Advanced Technologies. 7*(4), 273–280.

[178] Du, E., Li, M., Xu, B., Zhang, Y., Li, Z., Yu, X. & Fan, X. (2024). *Macromolecules, 57*(2), 672–681.

[179] Xiong, H., Zhang, L., Wu, Q., Zhang, H., Peng, Y., Zhao, L. & Wu, J. (2020). *Journal of Materials Chemistry A, 8*(46), 24645–24654.

[180] Shan, S., Wu, X., Lin, Y. & Zhang, A. (2022). *ACS Applied Polymer Materials, 4*(10), 7554–7563.

[181] Choi, C., Self, J. L., Okayama, Y., Levi, A. E., Gerst, M., Speros, J. C. & Bates, C. M. (2021). *Journal of the American Chemical Society, 143*(26), 9866–9871.

[182] Zha, H., Cheng, L., Wang, Z., Liu, C. & Hong, C. (2024). *Science China Chemistry, 1*–9.

[183] Liu, Z. & He, C. (2023). *Progress in Organic Coatings., 175*, 107371.

[184] Marinow, A., Katcharava, Z. & Binder, W. H. (2023). *Polymers, 15*(5), 1145.

[185] Luo, T., Lu, C., Qi, J., Wang, C., Chu, F. & Wang, J. (2024). *Chemical Engineering Journal, 479*, 147729.

[186] Moad, G., Rizzardo, E. & Thang, S. H. (2009). *The Australian Journal of Chemistry, 62*(11), 1402–1472.

[187] Liu, H. & Chung, H. (2016). *Macromolecules, 49*(19), 7246–7256.

[188] Yao, L., Rong, M. Z., Zhang, M. Q. & Yuan, Y. C. (2011). *Journal of Materials Chemistry, 21*(25), 9060–9065.

[189] Syrett, J. A., Mantovani, G., Barton, W. R., Price, D. & Haddleton, D. M. (2010). *Polymers Chemistry, 1*(1), 102–106.

[190] Mondal, P., Behera, P. K., Voit, B., Böhme, F. & Singha, N. K. (2020). *Macromolecul Mater Engineer, 305*(6), 2000142.

[191] Ding, M., Jiang, X., Zhang, L., Cheng, Z. & Zhu, X. (2015). *Macromolecular Rapid Communications, 36*(19), 1702–1721.

[192] Szczepaniak, G., Fu, L., Jafari, H., Kapil, K. & Matyjaszewski, K. (2021). *Accounts of Chemical Research, 54*(7), 1779–1790.

[193] Xu, X., Jerca, F. A., Jerca, V. V. & Hoogenboom, R. (2020). *Macromolecules, 53*(15), 6566–6575.

[194] Jiang, L., Griffiths, P., Balouet, J., Faure, T., Lyons, R., Fustin, C. A. & Baeza, G. P. (2022). *Macromolecules, 55*(10), 3936–3947.

[195] Lee, Y. B., Suslick, B. A., de Jong, D., Wilson, G. O., Moore, J. S., Sottos, N. R. & Braun, P. V. (2024). *Advances in Materials, 36*(11), 2309662.

[196] Naguib, M., Rashed, A. & Keddie, D. J. (2021). *Journal of Materials Science, 56*, 8900–8909.

[197] Keddie, D., Naguib, M. & Rashed, A. (2021). *Journal of Materials Science, 56*, 8900–8909.

[198] Romero-Sabat, G., Gago-Benedí, E., Rovira, J. J. R., González-Gálvez, D., Mateo, A., Medel, S. & Chivite, A. T. (2021). *Composites: Part A Applied Science and Manufacturing, 145*, 106335.

[199] Bielawski, C. W. & Grubbs, R. H. (2007). *Progress Polymer Science, 32*(1), 1–29.

[200] Szwarc, M. (1956). *Nature, 178*(4543), 1168–1169.

[201] Matyjaszewski, K. (1993). *Macromolecules, 26*(7), 1787–1788.

[202] Elling, B. R. & Xia, Y. (2015). *Journal of the American Chemical Society, 137*(31), 9922–9926.

[203] Wilson, G. O., Caruso, M. M., Reimer, N. T., White, S. R., Sottos, N. R. & Moore, J. S. (2008). *Chemistry Materials, 20*(10), 3288–3297.

[204] Freedman, B. R. & Mooney, D. J. (2019). *Advance Materials, 31*(19), 1806695.

[205] Barrere, F., Mahmood, T. A., De Groot, K. & Van Blitterswijk, C. A. (2008). *Materials Science and Engineering: R: Reports, 59*(1–6), 38–71.

[206] Diesendruck, C. E., Sottos, N. R., Moore, J. S. & White, S. R. (2015). *Angewandte Chemie International Edition, 54*(36), 10428–10447.

[207] Kahar, N. N. F. N. M. N., Osman, A. F., Alosime, E., Arsat, N., Azman, N. A. M., Syamsir, A. & Abdul Hamid, Z. A. (2021). *Polymers, 13*(8), 1194.

[208] Aïssa, B., Therriault, D., Haddad, E. & Jamroz, W. (2012). *Advance Material Science and Engineering, 2012*(1), 854203.

[209] Menikheim, S. D. & Lavik, E. B. (2020). *Wiley Interdisciplinary Reviews: Nanomedicine and Nanobiotechnology, 12*(6), e1641.

[210] Hansen, C. J., Wu, W., Toohey, K. S., Sottos, N. R., White, S. R. & Lewis, J. A. (2009). *Advances in Materials, 21*(41), 4143–4147.

[211] Norris, C. J., Meadway, G. J., O'Sullivan, M. J., Bond, I. P. & Trask, R. S. (2011). *Advanced Functional Materials, 21*(19), 3624–3633.

[212] Williams, K. A., Dreyer, D. R. & Bielawski, C. W. (2008). *MRS Bullettin, 33*(8), 759–765.

[213] Saba, S. A., Mousavi, M. P., Bühlmann, P. & Hillmyer, M. A. (2015). *Journal of the American Chemical Society, 137*(28), 8896–8899.

[214] Bobrin, V. A., Yao, Y., Shi, X., Xiu, Y., Zhang, J., Corrigan, N. & Boyer, C. (2022). *Nature Communications, 13*(1), 3577.

[215] Amaral, J. & Pasparakis, G. (2017). *Polymer Chemistry, 8*(42), 6464–6484.

[216] Bagheri, C. M. F. & Boyer, C. (2021). *Advanced Science*, *8*(5), 2003701.

[217] Tang, H., Luan, Y., Yang, L. & Sun, H. (2018). *Molecules*, *23*(11), 2870.

[218] Ekeocha, J., Ellingford, C., Pan, M., Wemyss, A. M., Bowen, C. & Wan, C. (2021). *Advances in Materials*, *33*(33), 2008052.

[219] Powell, P. C. & Housz, A. I. *Engineering with Polymers*. CRC Press, 2023.

[220] Elling, R., Su, J. K. & Xia, Y. (2020). *Accounts of Chemical Research*, 54(2), 356–365.

[221] Lee, J., Moon, H. H., Paeng, K. & Song, C. (2018). *Polymers*, *10*(10), 1173.

[222] Schubert, U. S., Winter, A. & Newkome, G. R. *Terpyridine-Based Materials: For Catalytic, Optoelectronic and Life Science Applications*. John Wiley & Sons, 2012.

[223] Destarac, M. (2018). *Polymer Chemistry*, *9*(40), 4947–4967.

[224] Keddie, J., Moad, G., Rizzardo, E. & Thang, S. H. (2012). *Macromolecules*, *45*(13), 5321–5342.

[225] Tanaka, K. & Matyjaszewski, K. (2008). *Macromolecules Symptoms*, *261*(1), 1–9.

[226] Krys, P., Schroeder, H., Buback, J., Buback, M. & Matyjaszewski, K. (2016). *Macromolecules*, *49*(20), 7793–7803.

[227] Cuthbert, J., Wanasinghe, S. V., Matyjaszewski, K. & Konkolewicz, D. (2021). *Macromolecules*, *54*(18), 8331–8340.

[228] Plaisted, T. A. & Nemat-Nasser, S. (2007). *Acta Material*, *55*(17), 5684–5696.

Chapter 6
Recent developments and potential application in self-healing polymers and elastomers

6.1 Introduction

Self-healing polymers and elastomers represent a growing area of materials science that aims to restore mechanical and structural integrity in response to damage. Recent developments in this field have focused on designing systems that can autonomously or externally respond to damage via molecular-level interactions. These systems incorporate dynamic covalent bonds, reversible supramolecular interactions, metal–ligand coordination, ionic associations, or embedded healing agents to enable repeatable and efficient repair cycles.

In particular, elastomeric materials, which are known for their flexibility and extensibility, have gained attention due to their ability to sustain damage under cyclic loading and yet potentially recover functionality when equipped with suitable healing mechanisms. Research in the past decade has led to the synthesis of elastomers capable of healing at room temperature or under mild external stimuli, such as heat, light, moisture, or mechanical force. These systems often exploit dynamic cross-linking networks – either through reversible Diels–Alder reactions, disulfide exchange, or hydrogen bonding – to initiate and complete healing.

The scope of self-healing polymers has also expanded beyond basic mechanical restoration to encompass multifunctional properties. For instance, conductive composites can now recover their electrical properties following fracture; shape-memory elastomers can restore both shape and strength; and hydrogels exhibit underwater healing through ionic cross-links or host–guest interactions. Additionally, newer fabrication approaches, such as microencapsulation, interpenetrating network design, and 3D printing have improved the integration of healing functionalities into polymer matrices.

The potential applications of self-healing polymers and elastomers are increasingly diverse. In structural engineering, these materials offer a promising route to prolong service life in load-bearing components. In electronics, they support the development of stretchable circuits and damage-tolerant sensors. In the biomedical field, soft self-healing materials are being explored for use in artificial tissues, wound dressings, and implantable devices due to their adaptability and biocompatibility. Furthermore, their use in coatings and sealants helps maintain barrier properties and aesthetics in consumer and industrial products.

This chapter outlines the key advances in synthesis, characterization, and functionality of self-healing polymeric systems, with a particular emphasis on elastomers. It also evaluates the performance metrics and technological readiness of these materials in real-world applications, while identifying critical challenges such as healing

https://doi.org/10.1515/9783111583716-006

speed, repeatability, mechanical robustness, and cost that remain to be addressed through future research.

6.2 Enhancement of mechanical properties

A major limitation of self-healing polymers is their inherently low mechanical strength and poor tensile properties, which often restrict their practical use in load-bearing or high-performance applications. To overcome these limitations, researchers have explored the incorporation of nanofillers – particularly graphene and its derivatives – into polymer matrices to enhance mechanical performance without compromising healing efficiency [1].

One such example is the work by Ding et al., who developed a double-network hydrogel composed of agar, polyacrylamide (PAM), and graphene oxide (GO) [2]. This system demonstrated remarkable fatigue resistance and outstanding self-healing ability. The initial mechanical characterization revealed a fracture strain as high as 4,600%, a fracture strength of 332 kPa, and an energy dissipation capacity of 11.5 $MJ \cdot m^{-3}$. After undergoing self-healing, the hydrogel retained significant mechanical integrity, with a post-healing fracture strain of 2000% and a fracture strength of 153 kPa. In this system, GO acted not merely as a passive filler, but as a dynamic cross-linker that interacted with the polyacrylamide chains, improving the overall tensile behavior through secondary interactions such as hydrogen bonding and electrostatic forces.

The mechanical reinforcement provided by graphene oxide is highly dependent on its dispersion state and concentration. At lower loadings (e.g., 0.5–1.0 wt%), GO sheets are uniformly distributed throughout the polymer matrix, leading to improved stress transfer and enhanced interfacial bonding. However, at higher concentrations (e.g., ≥3.5 wt%), GO sheets tend to aggregate due to strong π–π stacking and van der Waals interactions, which can lead to defects and reduced reinforcement efficiency [2].

In another study, Sun et al. synthesized a polyurethane (PU) composite reinforced with covalently bonded graphene oxide using a Diels–Alder (DA) reaction between furan- and maleimide-functionalized groups [3]. This reversible covalent chemistry not only endowed the material with a thermally activated self-healing function but also significantly improved the mechanical strength. The composite displayed an initial Young's modulus of 29.15 MPa and a fracture strain of 500%, both of which nearly doubled when the GO loading was increased from 0.1 wt% to 0.15 wt%. The enhanced performance was attributed to the homogeneous dispersion of GO and the strong covalent interactions with the PU matrix, which together facilitated efficient stress transfer and crack-bridging.

Furthermore, functionalizing the surface of graphene oxide with reactive groups improves its compatibility with the polymer matrix and enhances interfacial adhesion. Liu et al. fabricated a maleimide-functionalized graphene oxide/PU composite, in which the modified GO served as an effective cross-linking point within the matrix [4]. This system exhibited significantly improved tensile strength and fracture toughness. The maleimide groups facilitated covalent bonding with the polymer network and improved the dispersion of GO, which in turn reduced interfacial energy, increased matrix–filler interactions, and promoted energy dissipation during deformation.

Collectively, these studies suggest that the integration of graphene-based materials into self-healing polymers not only enhances mechanical integrity but also supports the healing function by maintaining structural coherence during and after damage. The combination of dynamic cross-linking chemistry and optimized filler dispersion remains critical in achieving multifunctional composites that offer both mechanical robustness and efficient self-repair.

6.3 Sustainable self-healing at ultra-low temperatures in structural composites

In self-healing composites impressive healing efficiencies can be achieved when circumstances are favorable. Under adverse conditions, healing might not be possible as observed in case of drastic low ambient temperature. In the report authors have demonstrated about structural composite, which is quite capable of maintaining its temperature to bring sustainable self-healing efficiency as seen in case of some animals who maintain a constant temperature to allow enzymes to remain active.

Certain laminates have been embedded with three-dimensional hollow vessels with the aim of transporting and discharging healing agents, and further a porous conductive material to provide heat internally to defrost and promote healing reactions. The authors have claimed to achieve a healing efficiency over 100% at about −60 °C [5–13].

Vaporization of sacrificial components (VaSC) and resin infusion have been utilized to fabricate fiber-reinforced composites (FRCs) incorporating hollow vessels and heating components. Manual embedment of poly (lactic acid) (PLA) conciliatory fibers such as 300 μm of VascTech fibers, has been incorporated in eight plies of woven glass fibers having area density of 290 gm^{-2} for respective ply in a square wave-like pattern. Layer-by-layer deposition of the reinforced fibers and remaining untreated reinforced fibers, and conductive plies in the particular sequence has been done [14–17].

In Fig. 6.1, two kinds of laminates such as random-discontinuous cotton-FRCs (type E with sub groups E1 and E2) having four sheets of cotton breather tissue, and woven carbon-FRCs (type C with sub groups C1 and C2) having four sheets of woven carbon fibers; have been tested out. E1 is random-discontinuous cotton string laminates and E2 is random-discontinuous cotton string laminates embedded with porous CNT layer; however C1 is woven carbon fiber laminates and C2 is woven carbon fiber laminates embedded with porous CNT ply [17–22].

The steady-state temperatures of the samples have been maintained in the range 20–85 °C as demonstrated in Fig 6.2 (a) after keeping voltage in the range 10–16 V. It is sufficiently high to enable healing in 24 h, as demonstrated in Fig 6.2 (b). As a result of the utilization of a low current such as 200 mA and its nice thermal conductivity,

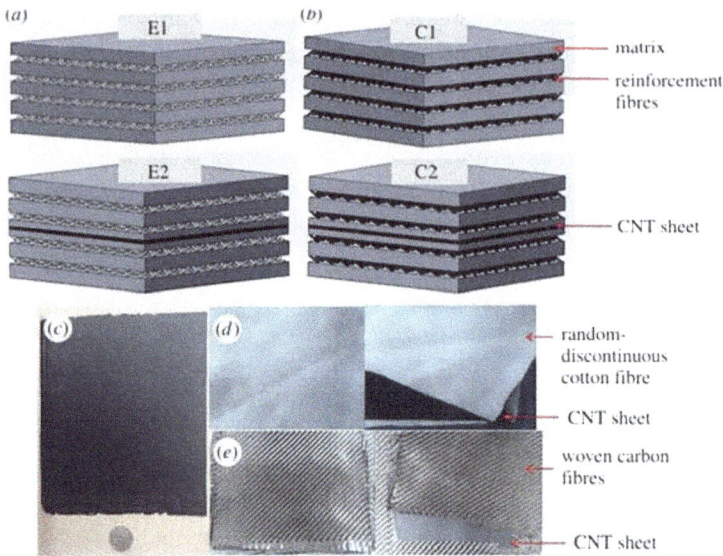

Fig. 6.1: (a) Demonstration of laminate E, (b) illustration of laminate C, (c) CNT porous ply, (d) random-discontinuous cotton fibers and CNT ply, and (e) woven carbon fibers and CNT ply [5].

there is no severe heat concentration. Fig 6.2 (c) shows that the laminate is able to be completely de-iced in 90 s and thereafter it is demonstrating the efficiency of carbon nanotube sheet (CNS) [23–26].

It is found that the laminate with copper foam sheet (CFS) is capable to maintain the temperature in the range 5–20 °C. It is also reported about design using CFS for 24 h is not long enough for it to heal completely, despite the healing agent being effective in the temperature range. One possible solution to this problem is raising the power of electricity to increase the interior temperature. That too is difficult as the copper foam has less resistivity. More electrical current of 55 A (in this case) is supposed to be required to generate the heat essential to keep the healing agent active in the way. After 24 h of healing at around −60 °C, the recovered mechanical properties of the samples are studied (as demonstrated in Fig. 6.3 (a, b)). At the end authors have concluded that for laminate with CNS, an average healing efficiency of 107.7% in terms of fracture energy and 96.22% in terms of peak load could be achieved. The maximum healing efficiency for fracture energy is found to be 141% (Fig. 6.3). In this way these results indicate that the laminates with CFS/CNS have capabilities to self-heal at ultra-low temperatures [27–33].

A few self-healing circumstances have been discussed in case of graphene/polymer laminates to restore initial properties.

Silicone rubber (SR) graphene nanoplatelets (GNPs) laminate has been manufactured, which can heal itself by thermal annealing, up to 250 °C in any oven for 2 h [6]. Also, researchers have fabricated ultrafast infrared (IR) laser-triggered self-healing lam-

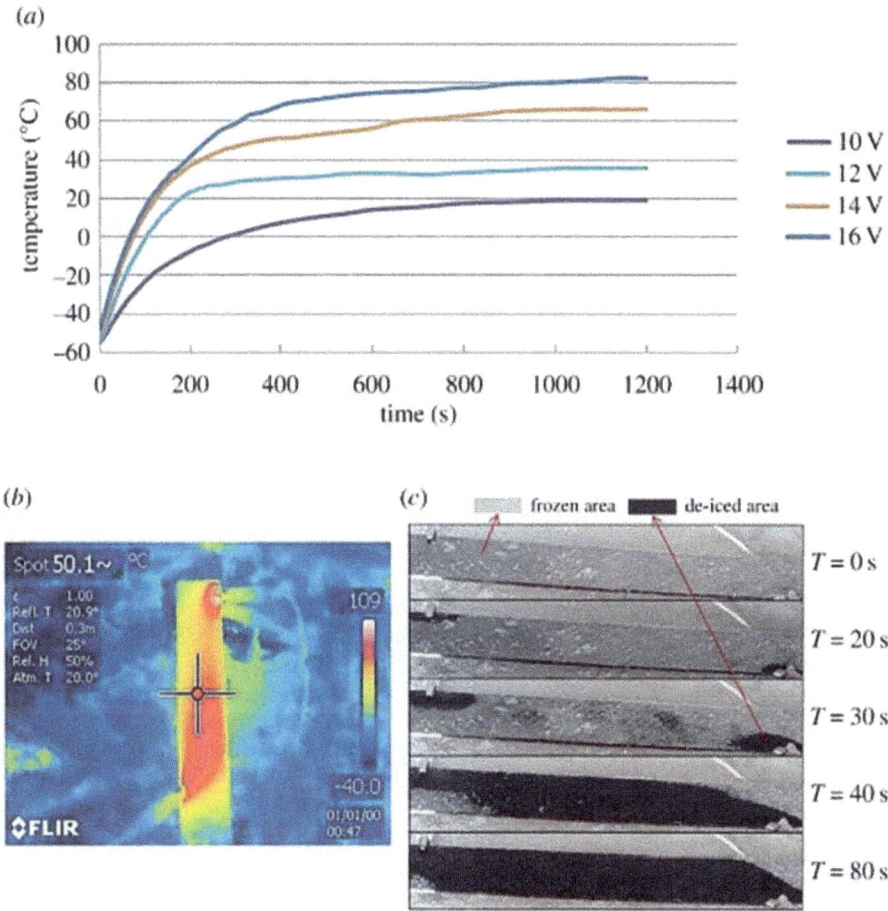

Fig. 6.2: (a) Temperature, sample in relation to time and voltage, (b) thermal distribution in the laminate baked by CNS in an ultra-low-temperature environment, and(c) de-icing laminate with CNS [5].

inate that can heal itself on increasing temperature from 30 °C to 150 °C in few seconds [7] as demonstrated in Fig. 6.4. As functionalized graphene nanosheets (FGNS) have IR absorbing capacity, using microwaves, self-healing can be accomplished as reported in case of covalently cross-linked reduced functionalized GO/PU laminates [35].

Self-healing is also achieved using other wavelengths of radiation as found in case of gold nanoparticles-reinforced poly(ε-caprolactone), which have been coated using RGO and silver nanowires [36–40]. Here, 91% improvement is achieved in terms of exterior conductivity and tensile strength.

Even self-healing can be accomplished at room temperature as reported in case of a mussel-inspired electroactive chitosan/GO laminate hydrogel and in this way it becomes more comprehensive [10–12]. A thermo-reversible elastomer (HBN-GO) has

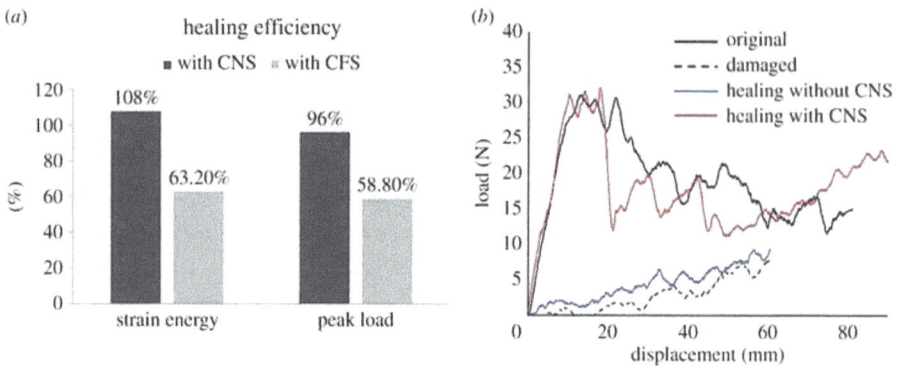

Fig. 6.3: (a) Healing efficiency of laminate having CNS and CFS, and (b) displacement–load relation for CNS specimen [5].

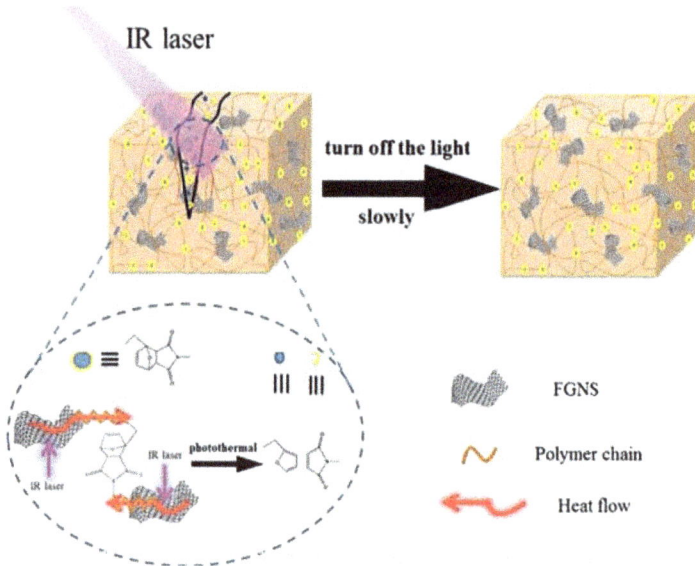

Fig. 6.4: FGNS nanocomposite is self-healed to 96% in terms of stiffness [7].

been synthesized and is reported to have efficient self-healing efficiency at room temperature without any enzymes [13].

Further, stress–strain curves have been plotted and tensile strength is examined for different specimens as demonstrated in Fig. 6.5.

Here, 60% improvement is achieved in terms of tensile strength, which is indeed remarkable improvement. Hydrogel based on β-cyclodextrin (β-CD) and N,N-dimethylacrylamide could heal itself at 37 °C and it has application as anticancer drug carrier [5]. Camptothecin (CPT) contents (as loaded and cumulated) when discharged in β-CD

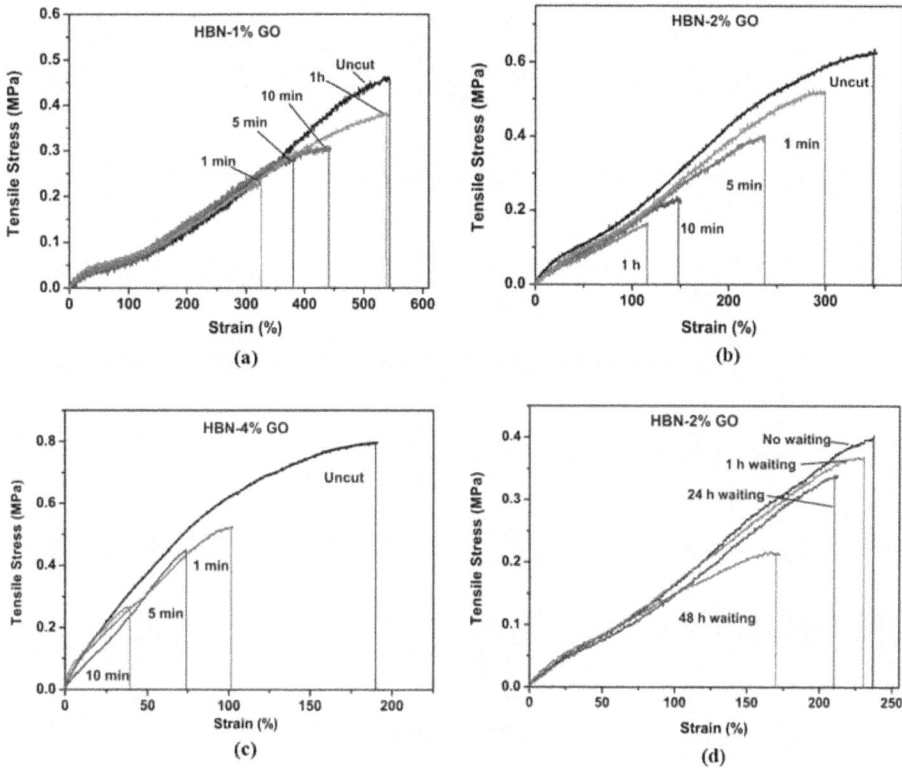

Fig. 6.5: Stress–strain curves of self-healing laminates (a) HBN – 1% GO; (b) HBN – 2% GO; (c) HBN – 4% GO at room temperature consequent to distinct healing time; (d) stress–strain curves of HBN – 2% GO consequent to 10 min healing time [9].

are found to be better than pristine graphene hydrogel as demonstrated in Fig. 6.6 (a). Self-healing hydrogels are quite effective while healing injury spontaneously [39] as illustrated in Fig. 6.6 (b, c).

Indeed it is a fact that self-healing composites have a bright future in the field of innovative product research. Many researchers are putting in efforts to recover functional properties in materials after healing the damages through these smart composites. Still, the field of self-healing composites has few limitations in understanding healing mechanism and thereby stability of healing functionality. Identification of damages and further healing are the main challenges for the self-healing composites.

After survey, it is impossible to ignore graphene/polymer laminates, which have self-healing capabilities in spite of many challenges in their use in practical applications. This is because improvement in self-healing capabilities and mechanical properties are two different conditions. Therefore, maintaining balance between these in a graphene/polymer laminate is still a challenging task. Also, it is equally essential to

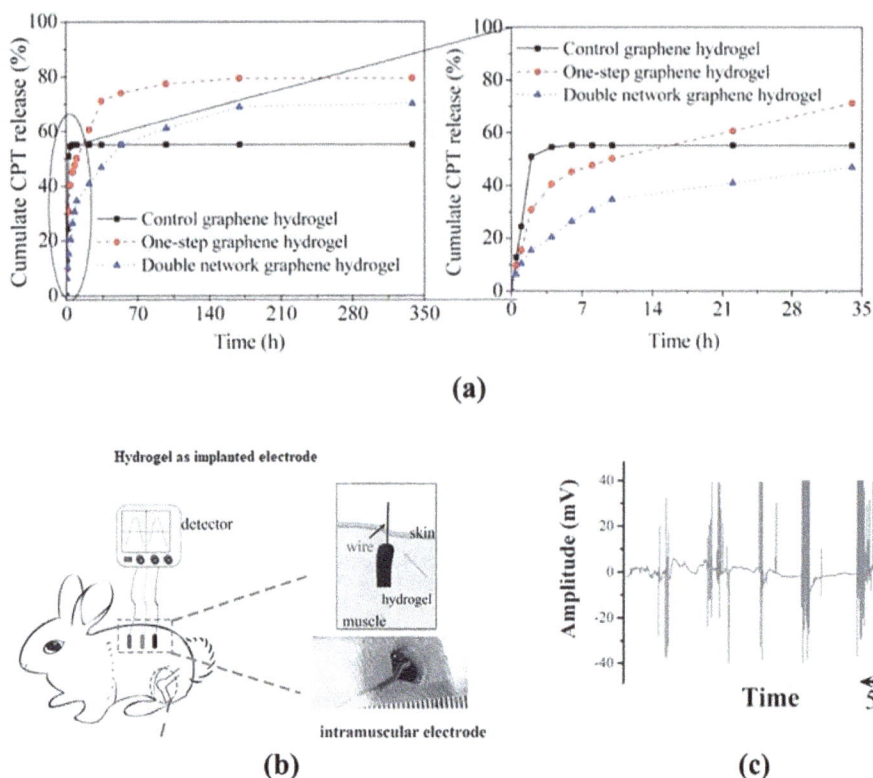

Fig. 6.6: (a) Polyborosiloxane (PBS) having discharged cumulate CPT [9]; (b) dorsal muscle having hydrogel electrodes implanted into it and electrodes are connected with sensors, which can detect signal; (c) illustration of recording of signal using hydrogel electrodes from rabbit's muscle when troubled [10].

improve compatibility of polymer and graphene after modification of graphene, while retaining natural properties of graphene as much as possible [39].

6.4 Development of shape memory effects

The word "memory" is derived from the Latin word "memor," which means remembering. Shape memory is a property of the material to return to its original position or conformation after deformation when subjected to stimuli. Shape memory materials (SMMs) can be classified as shape memory alloys, shape memory ceramics, shape memory polymer and gels. Shape memory alloys exhibit the memory effect by a phase transition from a high-temperature phase (austenite) to a low-temperature phase (martensite) [41]. Similar are the mechanisms for ceramics where zirconium dioxide plays a pivotal role in the crystalline polymer packing. Shape memory polymers (SMPs) work in a mechanism, which is very different from that of alloys or ceramics, mainly because of its glass transition temperature (T_g). In fact, SMPs are highlighted much more than alloys or ceramics because of their wider scope of mechanical diversity, lightweight, biocompatibility, and biodegradability.

i) Working principle from the thermodynamic perspective of shape memory polymers

a) We know that Boltzmann's equation relates the entropy to the conformation and is given as:

$$S = K \ln W$$

where S = entropy, K = Boltzmann constant, and W = probability of conformation

For an amorphous linear polymer chain, the entropy should be maximum, which in turn implies W to be maximum. Hence, the polymer takes the most coiled structure/conformation.

b) Polymers exhibit a special property of glass transition temperature (T_g), below which the polymer remains in a frozen state. The segmental motion begins as the temperature reaches T_g, and the phase is marked by a rubbery or elastic state. In this state, if a small force is applied to the polymer, it stretches in the direction of the applied force. Since the polymer is a highly coiled structure, a small force will result in a segmental deformation, but the neighboring chain entanglement will prevent further deformation [42]. Hence as the force is released, the deformed mass retracts back to its original position, which is the most probable state.

c) The entangled polymer may contain cross-linking points (which may be chemical or physical) that prevent the entire molecule from free flow. Thermoplastic elastomers contain these "net points." The maximum thermal transition (T_p) is correlated with the hard segment forming phase. If the temperature does not exceed T_p, the phases will stabilize the original shape by acting as physical net points. We know that a polymer cannot be crystallized completely, leaving behind amorphous parts in the system. If the glass transition temperature of this polymer is below the working temperature, the network will be elastic. They inherit a property called entropy elasticity [43], and as the polymer is trenched the loss of entropy increases. If the external force is released, the polymer reverts to its original shape regaining the lost entropy.

ii) Molecular perspective of shape-memory polymers

A polymer exhibiting memory effect has two segments – fixed phase and a switching phase. An elastomer will only exhibit this effect if it can be stabilized in the deformed state in a temperature range. For this, the consistency of the switching segment is so important. T_p is the temperature which is, in turn, a function of the switching segment. T_p can be either the glass transition temperature or the melting temperature, but usually melting temperature is preferred over the former. Above T_p, a strain-induced crystallization works within the switching system. As already mentioned, complete crystallization is not possible; hence the crystallization part acts as a fixed point, which prevents the chain from returning to its original conformation immediately. Thus a deformation is induced within the system. If the temperature is now raised beyond T_p of the switching segment, the segments are flexible and hence the original shape is retained.

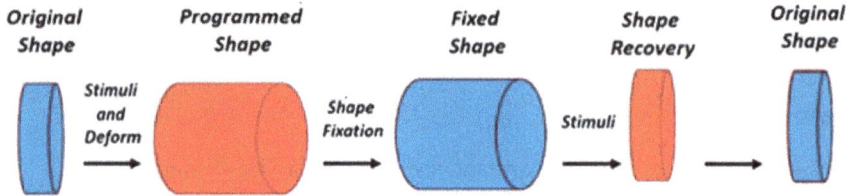

Fig. 6.7: To demonstrate shape memory effect in a polymer system.

iii) Areas of application

Shape memory elastomer finds its application in various research fields primarily because of its compatibility, wider tolerance range, easily processability and biodegradability. In the modern automobile industry, shape memory polymers are used as sensors and actuators for a safer and a tidier automobile with increased performances [44]. The electromagnetic actuators, which were used earlier caused several problems like flux damage, inaccurate resulting, field generation; but the introduction of drive-by-wire technology offers SMP actuators as an alternative. SMP is also used in aerospace hydraulic lines, which are resistant to high dynamic load and geometric space constraints. Artificial intelligence and neural networks have enhanced the usability of SMPs giving it space in robotic applications.

SMEs find a major application domain in biomedical science including aneurysm occlusion devices, vascular stents, and cardiac valve repair. The table below shows the major application field of SMEs in biomedical sciences.

Fields based on organs/tissues	Fields based on instruments
– Braces	– Catheters
– Palatal arches and files	– Ultra scope and endoscopy
– Head and spine	– Suture
– Bone	– Cardiology and hepatology
– Muscles	– Urology
– Fingers, hands, and legs	– Plastic and aesthetic surgery
– Aorta, arteries, and vessels	– Ophthalmology
– Vena cava filters	
– Ventricular septum	
– Valves (bicuspid and tricuspid)	

Despite the several advantages that the SMEs have, they fail to show stress recovery and sometimes the manufacturing is so costly that the ideation is not economical. A reduction in the strain recovery and an improved stress relaxation effect is desirable for many such applications. SMEs can be blended with nanofillers to improve the load-bearing capacities. Presently, endovascular protective filters use metal devices, which are placed inside vessels like the superior vena cava to prevent blood clots from traveling from the brain to the lungs. The problem arises when the clinical threats of the clot are minimized but the permanency of the metal filters possesses a threat of various side effects. SMPs are biodegradable in nature and if we can incorporate the same as an endovascular protective filter, it will turn out be a prodigious scientific invention.

Apart from the biomedical applications, SMEs can also impact the present aerospace components by forming a composite with a toughening agent, usually a metal like copper, aluminum or nickel. A composite, which will yield a number of cycles per operation along with a narrow loading hysteresis, can be evolved from SMEs [45, 46].

SMEs can also be used as a dampening or noise reduction agent in architectural sciences. In the field of artificial intelligence and robotics, SMPs can still explore their potential in micro and fast actuators, efficient and stable actuators, and rotary actuators. Self-healable composites can also drive the perspective towards coatings and brake pads. The concoction of mini- and microactuators such as NiTi films can be made possible with new cutting-edge technologies, which can further enhance SMPs attributes and functionality.

Although rated high in their mechanical properties, self-healing composites also display and exhibit the property of shape memory. This particular property is a lucrative feature in the context of polymer composites because of their wide applications in machinery, electronic device, chemical, and biological industry [47]. Weng and Dai et al. fabricated a series of graphene (polyacrylamide co-acrylic acid) composite materials with shape memory behavior complemented by self-healing property [48]. A primary unfold cube shape was developed for the validation section and was held heated at 35 °C. The compressed box possessed the ability to recover to its original position after being heated at 37 °C for 30 s and the cycle can be continued for at least 10 times. Another hyperbranched polymer fabricated by Karak et al., which comprised of the polyurethane titanium with reduced graphene oxide inhibited a wonderful shape memory property [49]. Being conducting in nature, the reduced graphene oxide absorbs the energy from the sunlight and judiciously passes on the absorbed energy to the polyurethane matrix. This results in the nanocomposite reaching its transition temperature much faster and hence the total number of cycle time increases. The surface temperature for recovering the shape was about 38.1–38.4 °C.

For conductive graphene/polymer composite, the shape memory property can be imparted by electrical conductivity. Joule heating is developed within the polymer matrix as soon as the content of the conductive filler exceeds the breakdown threshold of the system. Hence an important factor that regulates the cycle time of the shape memory process depends on the electrical conductivity of the filler.

Mohammad et al. molded a fast electroactive shape memory and self-healing PVA/graphene composite. The shape recovery experiments, which were initiated electrically, were conducted with 3 wt% and 4.5 wt % of graphene under four different voltages of (40, 50, 60, 70) V [50]. The outcomes were fascinating as increasing the graphene content from 3% to 4% leads to a drastic faster recovery response. The ratio of recovery rate and the graphene content also increased as the graphene content was increased from 3% to 4.5%

6.5 Advancement in rheological properties

Liu et al. had demonstrated the rheological properties of the self-healing polymers by plotting the dynamic storage (*G'*) and the loss (*G"*) with the frequency (ω), measured at different concentrations (*C*) after sufficiently long time intervals [51]. They found the gels to be developed with considerable mechanical properties at a concentration greater than 4.2 wt%. Further, they found it to exhibit a ω independent dynamic storage modulus coupled with a negligible factor of the loss modulus (*G"*). The frequencies of these gels exhibiting self-healing properties were equated to an equilibrium modulus (*G_e*).

The gel samples, which were stored at ambient temperature for a few minutes, were seen to flow, presumably by their own weight, which implies that these polymeric materials were analogous to dynamic covalent gels having a long dissociation time.

As the concentration of the system is reduced to less than 3.2 wt%, the system predominantly behaves as a viscoelastic fluid with explicit relaxation time. This brings us to the conclusion that the polymeric network may not have been developed completely at the low concentrations below 3.2 wt%. At a concentration of 3.7 wt%, *G'* and *G"* shows the power law behavior characteristic of the polymer exhibiting a critical gel condition. The self-healing polymers are generally embedded within these zones of concentrations displaying the following mathematical correlation [52–56].

In the pregel conditions,

$$\eta = \varepsilon^{-\gamma}$$

For the critical conditions as defined by Liu and his coworkers,

$$G' \sim G'' \sim \omega^n,$$
$$G''(\omega)/G'(\omega) = \tan \delta = \tan (n\Pi/2),$$

And, for the postgel conditions,

$$G_e \propto \varepsilon^z,$$

where η_0 is the zero shear viscosity, ε is the relative distance of a variable *q* from the gel point q_g

$$\varepsilon = |q - q_g|/q_g$$

where q is the dependency factor on the extent of the reaction, the gelation time, and the gelator concentration and the variables γ, n, and z are the scaling parameters used for the three power laws in the respective phases.

In the simplest level of the interactions, these parameters tend to follow am empirical relationship given by

$$n = z/(z + \gamma)$$

Researches focusing on the gelator concentration c and examining the scaling relationships, it can be demonstrated that the exact value of the critical concentration c_g can be evaluated by determining the loss tangent (tan δ). By plotting a graph according to the Winter Chambon equation, it is seen that tan δ is independent of the frequency (ω) [54].

The samples used by Liu et al. were exposed to the steady flow measurements, which in turn showed a zero shear viscosity at a concentration less than c_g. They further plotted a log-log graph of η_0 against ε, to predict the value of γ, which turned out to be 1.5. By using this value in the pregel phase, they calculated the value of z, which turned out to be 2.5. Both the exponents $\gamma = 1.5$ and $z = 2.5$ have overlapped with the percolation model as drafted by Martin et al. showing the correlation between the mathematical model and the practical behavior of the polymer system [57, 58].

If the similar model is projected for the polymeric system as devised by Liu et al., the predicted value of n turns out to be 0.63 against the observed value of n equal to 0.75. As the observed value of n is higher than the predicted theoretical value, we can infer that the material exhibited dynamic covalent nature. The same can either relax by the molecular motion of the gel network or by the thermal dissociation of the bonds.

6.6 Advancement in electrical properties

Polymers, in their native form, are typically electrically insulating materials. This is primarily due to the chemical structure of their backbone, which is commonly composed of covalently bonded carbon and hydrogen atoms. These covalent bonds are characterized by tightly bound, localized electrons that are not free to move across the material. Because there are no delocalized electrons or conduction bands, as found in metals or certain inorganic semiconductors, electron transport is not feasible within pure polymers under standard conditions.

Additionally, polymer chains are held together by weak intermolecular interactions, primarily van der Waals forces. These forces do not allow for strong orbital overlap between adjacent chains. The spatial separation between individual polymer molecules is typically large enough that interchain electron tunneling or hopping is severely restricted. Consequently, even if a localized charge carrier (such as a radical ion) were generated on one segment of the polymer, it would have great difficulty migrating across the bulk material, leading to extremely low electrical conductivity.

To overcome this limitation and impart electrical conductivity, conductive nano-fillers such as graphene or reduced graphene oxide (RGO) are incorporated into the polymer matrix. Graphene, a single layer of sp^2-hybridized carbon atoms arranged in a hexagonal lattice, exhibits high intrinsic electrical conductivity due to its π-electron system that allows for efficient electron delocalization across its surface. When embedded in a polymer, graphene creates conductive pathways or percolation networks within the otherwise insulating matrix. These networks act as bridges that facilitate charge transport through the composite.

The extent of conductivity enhancement depends significantly on the filler loading. At low graphene concentrations, the conductive sheets remain isolated within the polymer matrix, and charge transport remains limited. As the concentration increases beyond a certain percolation threshold, the graphene flakes begin to overlap and form continuous pathways. At this stage, a dramatic increase in electrical conductivity is observed, as electrons can travel through interconnected graphene domains. This transition is often nonlinear and depends on several factors, including filler morphology, dispersion quality, and polymer-filler interfacial interactions.

For instance, Geo et al. reported the fabrication of a multilayered polyelectrolyte film consisting of lithium polyethylenimine (Li-PEI), polyacrylic acid (PAA), and graphene. This film exhibited a charge transfer resistance of approximately 750 ohms, which represented a significant improvement over the resistance of the unmodified polymer blend. The reduced resistance was attributed to the creation of conductive pathways by graphene sheets embedded between the polyelectrolyte layers, facilitating more efficient ion and electron transfer during operation [60].

In the case of conductive hydrogels, water content becomes an additional and critical factor in determining the overall conductivity of the material. Water acts as a medium for ionic conduction, allowing for the movement of solvated ions such as Na^+ or K^+. Liu et al. synthesized a composite hydrogel using a superabsorbent polymer matrix, a hyperbranched polymer, and reduced graphene oxide via a hydrothermal synthesis method. This hydrogel demonstrated both high water absorption capacity and electrical self-healing behavior [61]. The key factor contributing to the electrical conductivity was the free water content, which enhanced the mobility of ions through the gel. The RGO nanosheets further improved the conductivity by enhancing ion dispersion and transport. Their large surface area and partial hydrophilicity allowed them to interact effectively with water molecules and dissolved ions, increasing the ionic conductivity of the hydrogel system [62].

Such graphene-based conductive self-healing hydrogels have become increasingly relevant in the design of flexible energy storage devices, especially supercapacitors. These devices require electrode materials that are both conductive and mechanically compliant. Liu and colleagues developed a stretchable supercapacitor using two parallel fiber springs coated with RGO. The fibers were further wrapped with a gel electrolyte and encapsulated with polyurethane. This structure provided both electrical connectivity and mechanical resilience, enabling the supercapacitor to stretch and re-cover its performance after mechanical deformation [61].

However, maintaining performance at low temperatures remains a persistent challenge for such systems. At sub-zero conditions, water in the hydrogel tends to freeze, which severely limits ion mobility and disrupts the electrochemical perfor-mance of the device. To address this, Chung and Ok designed a supercapacitor that could operate effectively at −30 °C by using a polyampholyte hydrogel as the electro-lyte and biochar-RGO composite electrodes [62]. The polyampholyte hydrogel pos-sessed hydrophilic polymer chains that strongly adsorbed water molecules, prevent-ing them from forming ice crystals. This adsorption suppressed the phase transition of water into ice, thereby preserving the ionic conduction pathways even at sub-zero temperatures. As a result, the supercapacitor retained its ability to store and dis-charge charge efficiently under cold conditions.

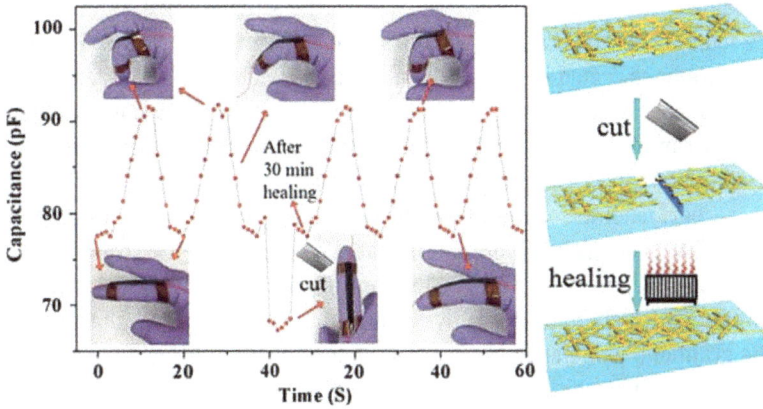

Fig. 6.8: Effect in the conductance in a self healing composite [112].

6.7 Anti-corrosion properties

To improve the durability and extend the service life of metallic substrates exposed to aggressive environments, the development of self-healing anticorrosive coatings has gained significant attention in recent years. These coatings are designed to autono-mously repair microscopic damage such as cracks or delaminations that occur due to mechanical, thermal, or chemical stresses. When damage occurs, the self-healing com-ponent embedded within the coating activates in response to external stimuli, such as heat or moisture, and restores the protective barrier, thereby preventing the initia-tion and propagation of corrosion at the exposed sites [63, 65].

One approach involves embedding nanostructured materials within polymer ma-trices to achieve both corrosion resistance and self-healing capability. For instance,

Lu and coworkers developed a nanocomposite coating composed of hydrated polyure-thane and graphene nanosheets modified with lignin. In this system, lignin was used as a modifier due to its high phenolic content, which promotes interfacial bonding between the polymer and graphene sheets, improving the dispersion of the filler and enhancing the mechanical integrity of the coating. Graphene, known for its barrier properties and high electrical conductivity, not only contributes to corrosion protection by physically blocking moisture and oxygen diffusion but also adds antistatic properties and ultraviolet radiation resistance. These combined features resulted in a coating that performs multiple functions, including restoring structural continuity, preventing electrochemical reactions, and resisting environmental degradation [65].

In another study, Goo et al. fabricated a self-healing composite by dispersing graphene oxide microcapsules (GOMCs) within a polyurethane matrix. The formulation was applied to a hot-dip galvanized steel (HDG) substrate, which is often used in structural applications requiring corrosion protection. The anticorrosive performance of the coated substrate was evaluated using a salt spray test, wherein the samples were continuously exposed to a 5% sodium chloride (NaCl) solution to simulate marine or deicing conditions. The standard polyurethane-coated steel plate began to show evidence of degradation, including microcracks, corrosion products, and delamination within a few days of exposure. In contrast, the sample coated with the GOMC-infused polyurethane showed no visible signs of corrosion or fatigue, even after five days of salt spray testing [63].

The enhanced performance of the GOMC/PU system can be attributed to several factors. First, the GOMCs act as nanocontainers that can encapsulate corrosion inhibitors or healing agents. Upon crack formation, these microcapsules rupture and release their contents, which chemically or physically seal the crack. Second, the layered structure of the graphene oxide sheets increases the tortuosity of diffusion pathways for corrosive species such as chloride ions and oxygen molecules, significantly reducing their penetration rate through the coating. Third, the presence of graphene oxide introduces additional cross-linking sites within the polyurethane matrix, enhancing the mechanical strength and thermal stability of the coating.

The process of healing was further investigated by thermal treatment. Samples that were intentionally fractured and then exposed to elevated temperatures exhibited partial or full reformation of the damaged interface. This thermally triggered healing was particularly evident in coatings containing GOMCs, where the fracture line gradually diminished upon heating, indicating material flow and rebonding across the damaged zone. In contrast, neat polyurethane coatings lacking these nanocapsules showed no such recovery, and the cracks remained visible and unaltered even after thermal exposure [65].

These findings demonstrate that the integration of graphene-based nanostructures into polymer coatings can significantly enhance both the passive and active protective mechanisms against corrosion. Such materials not only extend the lifespan of metallic components under harsh service conditions but also reduce the need for frequent maintenance or recoating. The dual role of graphene in both mechanical reinforcement and electrochemical shielding, along with its ability to carry and release healing agents, makes it a versatile and effective additive in the design of smart, self-healing coating systems.

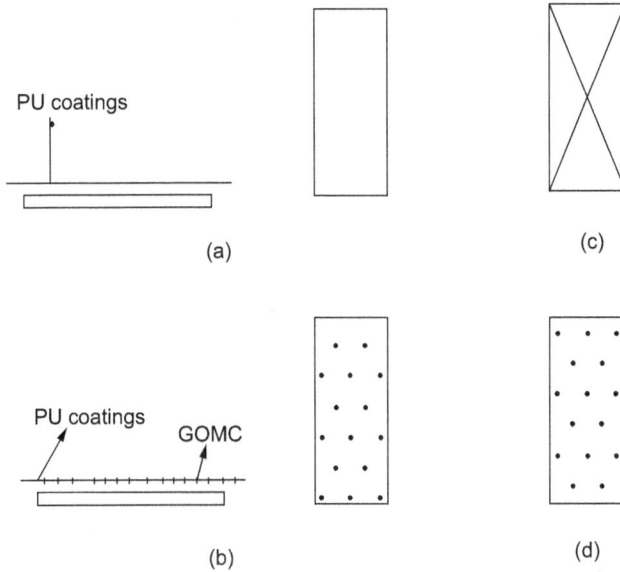

Fig. 6.9: To demonstrate the self healing properties in coating applications using polyurethane [65].

6.8 Biological and pharmaceutical application

In recent years, there has been significant interest in the biological applications of graphene-based self-healing materials, primarily due to their unique combination of electrical conductivity, mechanical robustness, and tissue compatibility. A large proportion of these materials is based on hydrogel systems, which are known for their high water content, flexibility, and intrinsic biocompatibility. This biocompatibility allows hydrogels to integrate with biological tissue with minimal immune response, making them suitable for implantation and interaction with living systems [31–33, 35, 38, 39, 66–69].

In addition to hydrogels, organogels and other soft composite materials have also been explored for their potential in biomedical applications. Organogels, which are composed of an organic liquid phase immobilized within a polymeric network, provide tunable mechanical and chemical properties that can be matched to specific tissue types [37]. Other advanced materials combining polymeric backbones with bioactive or conductive fillers have also emerged, expanding the design space for bio-integrated devices [40].

One notable example is the hydrogel developed by Wang K. et al. [33], which was engineered by partially reducing graphene oxide (GO) into conductive graphene using polydopamine (PDA) as a reducing and adhesive agent. The resulting composite, referred to as PDA-pGO-PAM hydrogel, is electrically conductive, stretchable, self-adhesive, and exhibits autonomous healing when damaged. Due to these combined features, the hydrogel is suitable for use as an implantable electrode material for recording in vivo bioelectrical signals. Upon implantation into the dorsal muscle of a rabbit, the hydrogel provided stable and high-quality electrical signals, demonstrating its functional integration with muscle tissue.

Further histological analysis confirmed that the hydrogel established close physical contact with the surrounding muscle fibers without triggering a detectable immune response [33]. This was evidenced by the absence of inflammatory cell infiltration or fibrous encapsulation in the adjacent tissue. More impressively, in a separate application involving cartilage regeneration, the same hydrogel material supported the formation of a continuous and mature cartilage layer over the defect site, indicating that it promoted tissue remodeling and integration over a longer duration. These findings emphasize that the hydrogel not only functions effectively as a biosensor but also supports tissue regeneration, highlighting its potential in both diagnostic and therapeutic applications.

One of the major challenges in any biological or pharmaceutical products is the consciousness of microbial contamination or infection. Microbial degradation can undermine the superior properties of the self-healing polymer nanocomposite. Self-healing polymers exhibit a brilliant property of antimicrobial action. An illustrative example of the same can be provided by the nanocomposites fabricated by Karak et al., which demonstrated an excellent microbial resistant property against *Staphylococcus, Escherichia coli,* and *Candida albicans.* The nanocomposites were composed of sulfur-containing groups along with polysulfones, which are resistant to the antimicrobial attacks [70].

The recent biomedical field has a great interest in self-healing polymer nanocomposites. Hu et al, synthesized a double network hydrogel based on β-CD functionalized graphene and *N, N*-dimethylacrylamide, which holds the capability of achieving the self-healing property along with bio-resistivity at 37 °C [71]. Before the secondary hydrogel network was introduced, a model anti cancer drug named Camptothecin(CPT) was impregnated into the primary hydrogel matrix. The loaded CPT and the secondary content CPT release in a β-CD functionalized graphene hydrogel showed better biocompatibility than that of the pure polymer.

Along with the above-discussed properties, the hydrogels do possess a diversifying biological and environmental applicability due to their compatibility with the DNA. A report generated by Shi et al. mentions the synthesis of GO/DNA self-healing hydrogels having a very high mechanical strength besides excellent environmental stability and high dye adsorption capability. These enhanced properties can be deployed in the application of gene therapy, drug delivery, tissue engineering, and removal of toxic pollutants [72].

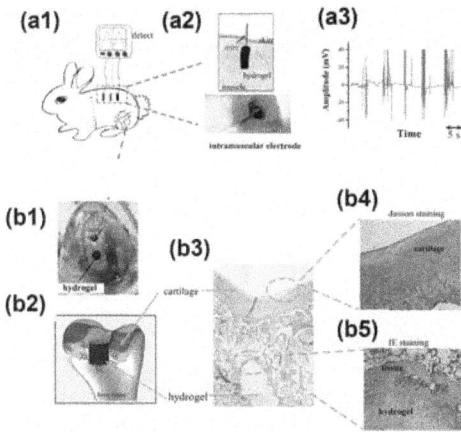

Fig. 6.10: The evaluation of PDA-pGO-PAM hydrogel. (a) The hydrogel as intramuscular electrodes. a1) Three hydrogel electrodes implanted into the dorsal muscle and the wires from the electrodes were transcutaneously connected to the signal detector. a2) Photos of the hydrogel implantation. a3) Example of EMG recorded by the implanted hydrogel electrodes from the muscle when the rabbit was interfered by external stimulation. (b) Long-term biocompatibility evaluation. b1) A cylindrical hydrogel with a diameter of 3.5 mm and a thickness of 5 mm was implanted in the osteochondral defect created at the femoral groove of the patellofemoral joint of a rabbit for 6 weeks, b2) Schematics of the hydrogel in the defect, which showed a blank from the hydrogel to the cartilage surface after surgery to allow tissue regeneration; b3) Histological images of the hydrogel treated defect (magnification × 40); b4) The amplified area showed the newly formed cartilage (magnification × 100); b5) The amplified area showed the intimate interface between the hydrogel and host tissue (magnification × 200) [33].

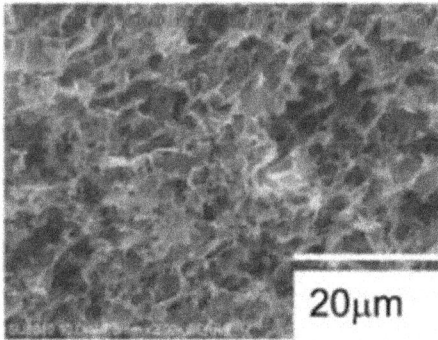

Fig. 6.11: SEM image of double network graphene hydrogel [194].

Self-healing hydrogels are so much widely used as a biomaterial because the property of healing the material is in as much less time as required. The mussel-shaped conductive and self-healing adhesive fabricated by Liu et al. had the potential to get implanted because of its excellent, long-term biocompatibility. Moreover, it showed

more accuracy in detecting a specific muscle in deep tissue apart from in vivo stimulating detection [73]. The hydrogels, mainly living intravascular electrodes resulted in
an excellent signal from the dorsal muscle after implantation into the deep muscle.
The range of the functional signals ranged 0.1–40 mV, which was more enhanced
than the signals emitted by the surface electrodes [74, 75].

6.9 Cutting-edge applications

6.9.1 Flexible electronic devices

Flexible electronic devices (e.g., flexible conductive devices, sensor, and artificial electric skin) attract much attention due to their unique ductility and their efficient and
low-cost manufacturing processes. However, they still faced poor mechanical performance and brittleness. Once damaged, it will be a failure and cannot be recovered,
influencing on the performance of the products, and even with a hidden danger.

Therefore, self-repairing flexible electronic devices with graphene and graphene
derivatives [88, 5, 33, 89–91–94] were expected to solve these problems. For example,
Wong C.-P. et al. [92] developed a covalently cross-linked reduced functionalized graphene oxide/polyurethane (RFGO/PU) composites based on Diels–Alder (D–A) chemistry and applied it in the healable flexible electronics. The composites show enhanced
mechanical property, good thermal stability, and excellent electromagnetic wave
healable property. Thereunto, 3D FGFs formed the conductive path in RFGO/PU composites that offered flexibility and stretchability (Fig. 6.12A). Moreover, the advantages
of the self-healing sensors are excellent mechanical properties, good conductivity,
self-healable capability, and sustainability [5, 33, 36, 89, 95, 90, 96]. For instance, C.-P.
Wong et al. [5] have reported a novel self-healing flexible sensor with three dimensional graphene structure with excellent stretchability (200%) and intrinsic conductivity, and highly effective healing performance by heat and microwave to detect human
motions. With the decrease of bending radius, R/R0 exhibited a slight increase during
the first bending cycle (Fig. 6.12 B-a). After the first cycle, the 3D FAGS/DAPU composite can almost recover the original value after straightening. Besides, the sensor exhibited the excellent stability of the resistance of FAGS/DAPUs after 1,000 cycles
(Fig 6.12B-b). In addition, the FAGS/DAPU composite was cut into slices with a width of
2 mm and connected by the copper wires. The as-prepared flexible strain sensor was
tied on the surface of the glove to detect the movements of the finger. In result, the as-
prepared strain sensor was able to detect the signals of the finger bending for cycles
(Fig. 6.12 B-c). Moreover, they investigated the durability of the FAGS/DAPU composites with a speed of 5 mm/min. After 200 cycles of stretching–relaxing under 10%
strain, the composites retained excellent stability (Fig. 6.12 B-d). The above descrip-
tions indicate that the FAGS/DAPU was suitable for flexible sensors. Last but important, artificial skin, possessing natural skin's pressure sensitivity and mechanical self-

healing properties, is a challenge for AI robots. Herein, Li Y. et al. [96] have described a thin film composed of graphene and polymers, holding both the mechanical self-healing and pressure sensitivity behavior of natural skin without any external power supply. Besides, its ultimate strain and tensile strength are even two and 10 times larger than the corresponding values of human skin, respectively. It also demonstrates highly stable sensitivity to a very light touch (0.02 kPa), even in bending or stretching states. Hence, it is a promising candidate for artificial intelligence skin.

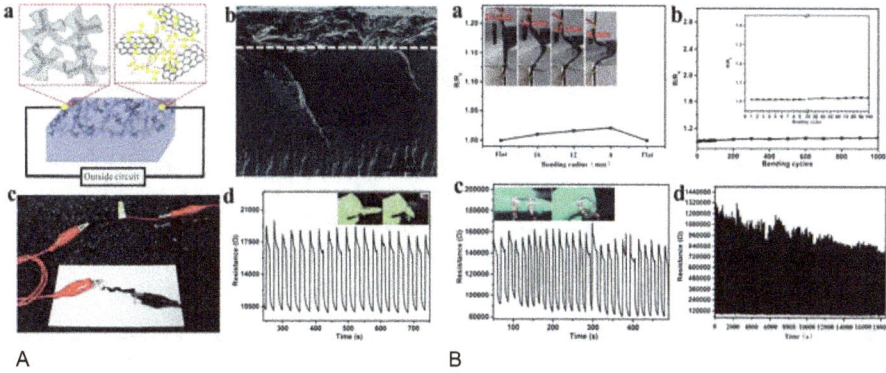

Fig. 6.12: A. The FGF/RFGO-DAPU-3 composite for flexible electronics. (a) Schematic diagram of flexible electronic circuit and compositions; (b) the cross-sectional image of the FGF/RFGODAPU-3 composite; (c) its application as a flexible conductor, and (d) a strain sensor. (Inset: corresponding digital photos of strain senor for the finger bending cycles) [58]. B. (a) R/R0 change of FAGS/DAPUs at a bend radius up to 8.0 mm in the first bending cycle. The inset pictures describe the bending process. (b) R/R0 of FAGS/DAPUs as a function of cycles for a bend radius of 8.0 mm. (c) Resistance-time curve of FAGS/DAPUs for the application of the flexible strain sensor during the finger bending cycles. (d) Durability test of the FAGS/DAPU as a strain sensor under 10% strain for 200 cycles [5].

6.9.2 Coating

Coating is generally used for protection, such as avoiding damage and anticorrosion. However, once the coating is damaged, the protection is a failure. Therefore, the development of self-healing coating is in great demand. Graphene-based self-healing materials as coatings [97–100] appeared at this time. For example, Gao S. et al. [98] prepared the graphene oxide microcapsules (GOMCs), containing linseed oil as the healing agent, embedded into waterborne polyurethane matrix, enabling the facile fabrication of self-healing composite coatings on hot-dip galvanized steel surfaces. In Fig. 6.13, the schematic drawings show that the primary GOMCs/PU composite coatings exhibited better anticorrosion properties, because the physical barrier effect of the GO shell elongated the pathway of the corrosive medium by restraining the diffusion of the small molecules (e.g., H_2O and O_2). When GOMCs/PU coatings were manually scratched with a scalpel blade and healed under ambient condition, and then subjected to the salt spray, no corrosion phenomena emerged on the HDG plate, because the micro-cracks in the composite coatings were healed autonomously, retarding the corrosion process. The above instructions show that the GOMCs/PU coatings have excellent self-healing ability and corrosion resistance.

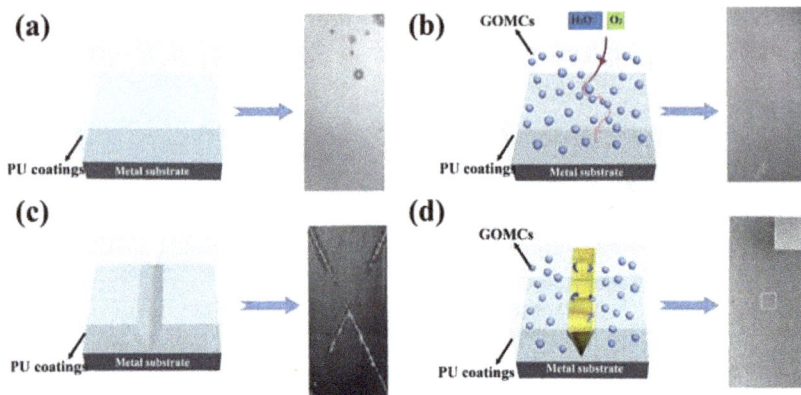

Fig. 6.13: Schematic and images. (a) Neat PU coating and (b) GOMCs/PU coatings, subjected to the salt spray test for 116 h. Schematic and images of (c) neat PU coating and (d) GOMCs/PU coatings after scratch and 15 days of healing, subjected to the salt spray test for 43 h. Inset: the enlarged view of the white block. Reprinted with permission from [98].

6.9.3 Supercapacitor

Energy is a hot topic, and graphene-based self-healing materials for supercapacitors [7, 101–102] have been reported in recent years. Zou Z. et al. [101] have reported reduced graphene oxide fiber-based springs as electrodes by wrapping fiber springs with a self-healing polymer outer shell for stretchable and self-healable supercapacitors, with the performances depicted in Fig. 6.14. Firstly, the supercapacitor can be easily stretched up to 100% (Fig. 6.14a). The CV curves had little shrinkage of enclosed area when the supercapacitor was stretched from 0% to 100% (Fig. 6.14b). At a stretch of 100%, the device retained 82.4% capacitance retention (Fig. 6.14c). After being cut and healed, the capacitance of the device was still maintained at a high level (Fig. 6.14d).

The GCD results (Fig. 6.14e), consistent with those in Fig. 6.14d, show good restoration of the supercapacitor. For better understanding, the EIS of the stretchable and self-healable supercapacitor was investigated. The ESRs of the capacitors before cutting and after the first, second, and third healing are 55.2, 56.2, 61.9, and 81.4 Ω, respectively. The first and second impedance spectra nearly mixed together (Fig. 6.14f), due to good reconnection of the electrode. In addition, the stretchable and self-healable supercapacitor was used to drive a photodetector of perovskite nanowires (Fig. 6.14g).

Before cutting and after healing (Fig. 6.14h), the on/off ratio only had a slight decay (Fig. 6.14i), which demonstrated the good restoration of the supercapacitor's function. This work gave an essential strategy for designing and fabricating stretchable and self-healable supercapacitors in next-generation multifunctional power devices.

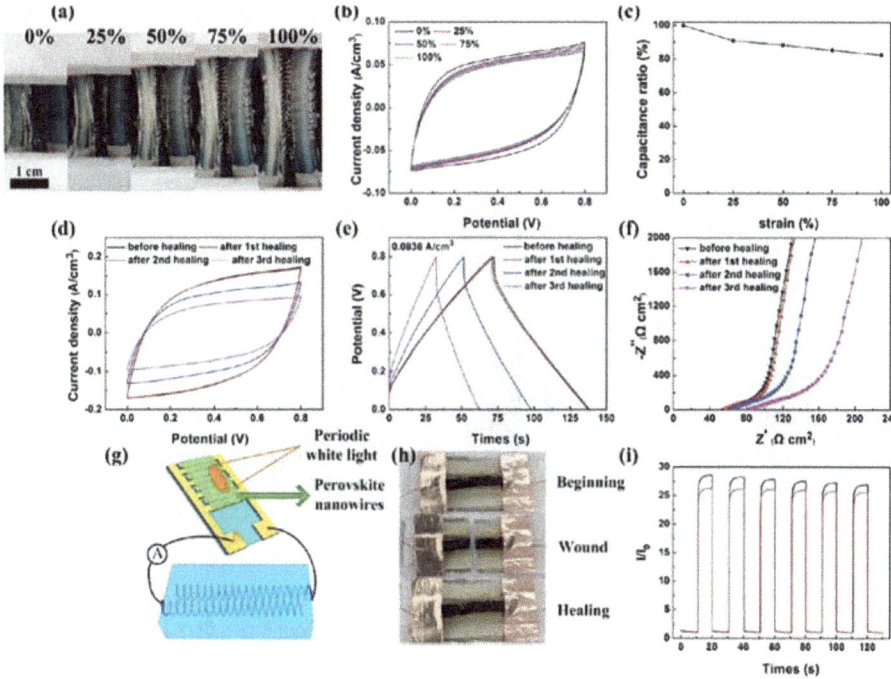

Fig. 6.14: Electrochemical measurements and application for as-prepared stretchable and selfhealing supercapacitors. (a) Photographs, (b) cyclic voltammo gram curves, and (c) evolutions of specific capacitance of the supercapacitor before and after stretching to 100%. (d) Cyclic voltammogram curves, (e) galvanostatic charge–discharge measurements, and (f) Nyquist plots of the supercapacitor before healing and after several self-healings. (g) Illustration of the supercapacitor driving a photodetector of perovskite nanowires. (h) Photographs of the supercapacitor before and after self-healing. (i) Photocurrent dependence on time of the photodetector under illumination of on/off states driven by the original and self-healing supercapacitor after a healing cycle; red corresponds to the self-healing supercapacitor and black to the original [101].

6.9.4 Self-cleaning

The environment is the foundation of human existence, and it is a challenge to study self-healing materials with environmental protection function. Here, Karak N. et al. [103] fabricated the hyperbranched polyurethane (HPU)-TiO$_2$-reduced graphene oxide (TiO$_2$/rGO) nanocomposite, with sunlight-induced self-healing and self-cleaning properties by degrading the dyes with excellent degradation efficiency. During degradation, rGO absorbs the energy of sunlight and emits shorter wavelengths of light. The emitted energy excites the TiO_2 nanoparticles to generate $h+/e$__pairs, which react with H_2O and O_2 to form reactive oxygen species (ROS) such as •OH and •O_2^-. These

ROS actually degrade the dye molecules. TiO_2 nanoparticles in the nanocomposite are only responsible for degradation of the dye. Therefore, it can be used as a surface coating material where it can remove the presence of organic dirt and provide a clean surface (Fig. 6.15).

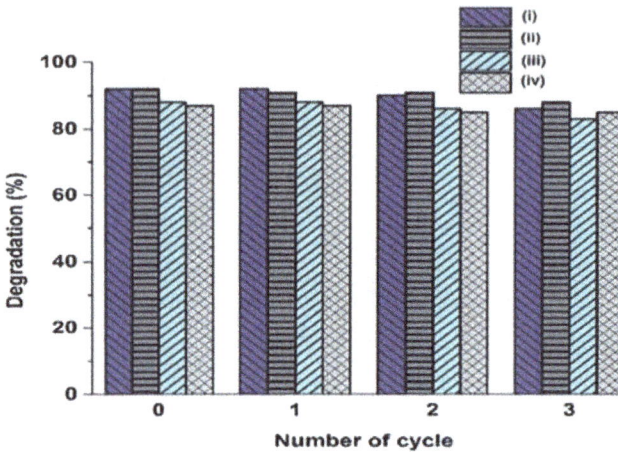

Fig. 6.15: Photocatalytic efficiency of (i) HPUT10RGO10, (ii) HPU-T10RGO5, (iii) HPU-T5RGO10, and (iv) HPU-T5RGO5 toward the degradation of MB in four separate cycles of the continuous dark-adsorption and photocatalytic processes [103].

6.9.5 Adsorbent

Wastewater treatment is an important environmental protection topic. Nevertheless, the cost is very high because of the irrepairable adsorbent material, which is an urgent problem to be solved. Ran R. et al. [104] prepared a new class of graphene oxide (GO)/hydrophobically associated polyacrylamide composite hydrogels (GHA gels), which exhibited prominent self-healing ability and fatigue resistance, especially the capability as recyclable adsorbents for dye wastewater treatment. The material has very high adsorption efficiency and cyclicity, and the introduction of GO significantly improved the purification ability of the HAPAM gel for the treatment of dye wastewater (Fig. 6.16).

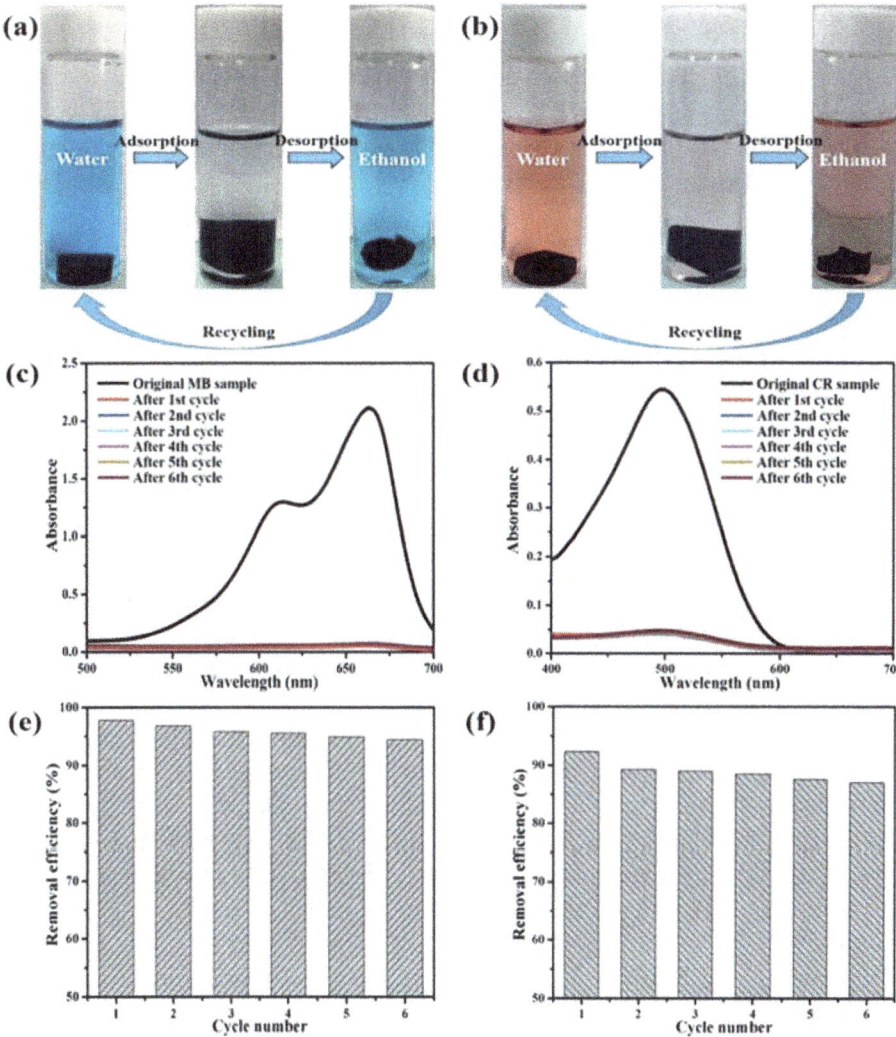

Fig. 6.16: Photographs manifesting typical cycles of purifying water and recovering adsorbent by G5HA for MB (a) and CR (b) solutions. UV/vis spectra of MB (c) and CR (d) solutions before and after cyclic adsorption. Removal efficiencies of MB (e) and CR (f) on G5HA during several adsorption–desorption cycles. The washing time in ethanol for each desorption process was fixed at 30 min [104].

6.9.6 Biomimetic materials and actuators

A novel bionic composite material inspired by the natural process of yeast fermentation was developed by Pugno NM and colleagues [36]. In this study, graphene nanoplatelets were introduced into a nutrient broth environment where yeast cells were growing. It was found that the graphene nanoplatelets were able to migrate and attach themselves to areas of the yeast cell surface that had developed stress-induced cracks. This reassembly effectively repaired these damaged regions on the cell surface. As a result, the composite material partially restored the original electrical conductivity and mechanical strength of the yeast cells that had been compromised by the stress. This method offers a unique approach to creating self-healing bioelectronic materials. The ability of the graphene to repair living cell surfaces points to potential applications in bioelectronic devices that can heal themselves, as well as in biosensors that rely on microorganisms to detect strains or chemical changes in the environment.

Actuators, which are devices that convert energy into motion, are a promising type of intelligent material mostly made from soft components [105–107]. These actuators are particularly useful in creating tiny robots or "microrobots" that can operate in small, confined, or enclosed spaces. However, such soft actuators are often prone to damage caused by external forces or environmental factors. Therefore, developing actuators that can repair themselves after being damaged is important for improving their durability and reliability.

Maganti LS and colleagues [108] addressed this challenge by synthesizing a stable nanogel made from easily produced graphene nanoparticles combined with polyethylene glycol. This nanogel displays a shear-thinning behavior, meaning its viscosity decreases when a force is applied, which is useful for flexible movement. The graphene nanoflakes within the gel interact and weave between the polymer chains, likely due to a refluxing process during synthesis. This creates a network structure that behaves like a combination of springs and dampers, allowing the gel to absorb stress and deformation. Notably, this nanogel can heal itself after damage without needing any external triggers, regaining its original elastic and mechanical properties. The team also developed a mathematical model describing the gel's viscoelastic behavior, showing that it can be represented as a modified Burger's model, which explains materials that exhibit both elastic (spring-like) and plastic (permanent deformation) behaviors.

Due to these properties, this graphene-based nanogel shows strong potential for use in microscale actuators, which require materials that are flexible, durable, and capable of self-repair in confined and delicate environments.

6.10 Future scope

As we have already mentioned, self-healing graphene/polymer composite is one of the bolstered smart materials, which will definitely make a severe impact on the technological innovation in the future years. The latest research, progress, and the innovation are concise in Table 6.1. In spite of the advantages we have achieved using poly-

Tab. 6.1: Properties of various self-healing graphene polymer composite.

Material	Self-healing mechanism	Self-healing condition	Self-healing efficiency				Mechanical properties			Application	Reference
			Electrical condition	TensilsStress	Tensile strength	Toughness	Tensile strength	Young's modulus	Toughness		
1.FG/TPU	Diffusion (simple chain diffusion)	IR, UV, microwave	98%				40 MPa			Electronics, construction engg.	[76]
2.GO/PAA hydrogels	Diffusion & hydrogen bond	Thermal		88%	92%	96%	0.35 MPa			Biomedical	[77]
3.GO/PU	Covalent bonding/	Thermal		76%	72%		22 MPa	22 MPa	21.95 MPa	Smart materials	[78]
4.RFGO/PU	Diels–Alder	Microwave		93%	93%	98%	24 MPa Elongation = 700%			Actuators	[79]
5.GO clay PDMMA hybrid hydrogel	Diffusion	Near-infrared		96%	94%	92%	184KPa Elongation = 1890%			Smart structure material	[80]
6.Superabsorbent hydrogel, hyper branched polymer/RGO	Diffusion	hydrothermal	98%				48 MPa			Biomedicine	[81]
7.Epoxy resin/Graphene	Diels–Alder	Microwave		96%	94%	94%	68% (of the initial)			Anticorrosive coatings	[82]

(continued)

Tab. 6.1 (continued)

Material	Self-healing mechanism	Self-healing condition	Self-healing efficiency				Mechanical properties			Application	Reference
			Electrical condition	TensilsStress	Tensile strength	Toughness	Tensile strength	Young's modulus	Toughness		
8. PBS/GO	Dative bond	Simple contac.	Almost 100%	94%	92%	92%				Electronics	[83]
9. HBN/GO	Diels–Alder	Near infrared		88%	86%	86%	20.2 MPa	19.4 MPa		Actuator	[84]
10. GO Hectorite clay/poly N,N-dimethyl acrylamide	π–π stacking, dative bond	IR		96%	96%	96%	10.2 MPa			biomedicine	[85]
11. Polyethylene imine PEI	Reversible bonds	Thermal			87%		13.3 MPa*			Seals, Hose	[86]
12. SHPU/graphene	Interchain diffusion	IR	39%	84%	80%	80%	4 MPa	3.8 MPa		Functional polymers	[87]

mer nanocomposites as a self-healing polymer, there are various challenges to solve in the near future for a pragmatic outlook.

Graphene, which is a pivotal part of the entire self-healing process, can also play the role of self-healing composites due to its photothermal energy transformation capability. Considering this case, we may conclude that graphene acts as an energy absorber, which absorbs the radiation on the photon energy into thermal energy very efficiently to facilitate the mutual diffusion of the polymer chains across the fractured surfaces thus initiating the self-healing process. On the other hand, graphene can sometimes be used as the reinforcing agent to elevate the mechanical strength. Self-healing process, in this case, can be promoted due to the interaction between the graphene and the polymer, be it covalent bonding or π–π stacking, immobilizing the diffusion of the polymer chains. Graphene also has a prime role in considering the glass transition temperature of the polymer, which in turn induces the self-healing efficiency, but unfortunately, mechanical properties and the self-healing efficiency are two contrasting trails. Increasing the mechanical properties by incorporating graphene reduces self-healing efficiency. This still remains a puzzle to solve for the scientist on optimizing both the properties. Moreover, self-healing graphene-based composites need to heal intrinsically; the method of simple contacting without the supplement of any stimuli is preferred. Hence investigating more pathways for the simple contacting procedure is required.

Once again for conductive graphene/polymer composites, mechanical strength and conductivity are two opposing factors. Increasing the graphene content will improve the electrical conductivity but the mechanical properties fall. Optimization of these three parameters yielding an excellent overall property composite is desired and needs to be discovered. On an additional note, a variety of polymeric materials should be investigated, which will show self-healing properties. Although it is quite difficult to induce the property to all the polymers, research must be carried out to find ways of grafting polymers that can develop the property of self-healing.

Finally, graphene needs to be modified in order to compose the polymer and the graphene together. However, modification of graphene oxide involves a decrease in inherent property, which the original graphene possessed. Therefore improving the compatibility as well as retaining the inherent properties of graphene is still an area that needs to be rediscovered.

6.11 Conclusion

Although we have made remarkable progress in self-healing polymer composites, there are several constrictions till date, which make the job of engineers more challenging. But interdisciplinary studies coupled with collaboration amongst various scientists are trying to address the problem starting from inherent crack-repairing ability to fabricating novel material, which displays self-healing property.

Extrinsic self-healing processes are synchronized for large-scale use for various applications including the synthesis of intelligent materials, which further supplements reduced waste disposal. Now with the age of the digital science, computer simulation including probabilistic models and Big Data analysis can aid the process of self-healing. Lee et al. had fabricated a solid state device, which includes polymeric layers and graphene nanofillers [109]. They used computational modeling, which demonstrated that incorporating nanofillers in the polymer composites will cause the fillers to be localized in the cracked surface and will intrinsically heal the cracks via their chemical or physical mechanisms. Trau et al. used an electrohydrodynamic system in which the system simulated that the cracks can heal within 60 nanoseconds if exposed to an electric field [110]. Karger Kocsis had integrated both the properties of self-healing and shape memory by blending thermosets with thermoplastics [111]. In these smart materials, generally, the thermoplastic part provides the system with the switching and healing effects while the thermosets behave as the fixed points.

For long-term prospects and goals, the fabrication of these innovative smart materials supplemented by intrinsic self-healing through various molecular designs is desired. However, although several pieces of research have shown the initiating trend in this field, the automated trigger mechanism still remains to be exposed. Finding an answer to these assertions accompanied by designing the solution would definitely push the engineering aspect of polymer science far ahead than its recent advancements.

Keeping these in consideration, we have discussed the recent progress and advances in graphene-incorporated polymers exhibiting self-healing property. The preparation method, conditions that favor self-healing properties and advantages along with future scope have been reviewed. Smart materials, especially self-healing materials are still in their genesis status, which will surely advance within the next few years. There will be a huge exploration in this field, which will disrupt the entire biomedical and the electronics technology [ref. from Sayan's thesis].

References

[1]   Kong, J. & Lu, X. (2017). Aqueousonly, green route to self healable UV resistant PU/ graphene lignin nano composite coatings. *ACS Nanotechnology, 11.* 2066–2074.

[2]   Ding, Y. & Hu, M. (2017). One pot fabrication of a novel agar polyacrylamide/ graphene oxide nanocomposites double networked hydrogel with high mechanical properties. *ACS Applied Materials & Interfaces, 9.* 38052–38061.

[3]   Sun, R., Jiang, K. & Ding, L. (2014). In situ polymerization of mechanically reinforced, thermally healable graphene oxide/ polyurethane composites based on Diels Alder Chemistry. *Journal of Materials Chemistry A, 2.* 20642–20649.

[4]   Liu, X. & Xu, J. (2017). A self healable nano composite based on dual crosslinked GO/. *PU Polymer, 127.* 241–250.

[5]   Li J, Liu Q, Ho D, Zhao S, Wu S, Ling L, Han F, Wu X, Zhang G, Sun R, Wong CP. Three-Dimensional Graphene Structure for Healable Flexible Electronics Based on Diels–Alder Chemistry. ACS Appl Mater Interfaces. 2018 Mar 21;10(11):9727–9735. http://doi:10.1021/acsami.7b19649. Epub 2018 Mar 8. PMID: 29436214.

[6]   Yanagisawa, Y., Nan, Y., Okuro, K. & Aida, T. (2018). Mechanically robust, readily repairablepolymers via tailored noncovalent cross-linking. *Science, 359*(6371), 72–76.

[7]   Yue, Y., Liu, N., Ma, Y., Wang, S., Liu, W., Luo, C. *et al.* (2018). Highly Self-Healable 3D Microsupercapacitor with MXene-Graphene Composite Aerogel. *ACS Nanotechnology, 12*(5), 4224–4232.

[8]   White, S. R., Sottos, N. R., Geubelle, P. H., Moore, J. S., Kessler, M. R., Sriram, S. R. et al. (2001). Autonomic healing of polymer composites. *Nature, 409*(6822), 794–797.

[9]   Toohey, K. S., Sottos, N. R., Lewis, J. A., Moore, J. S. & White, S. R. (2007). Self-healing materials with microvascular networks. *Nature Materials, 6*(8), 581–585.

[10]   Y., G., M, C., B, W., Z, F. & F, S. (2010). Diving-surfacing cycle within a stimulus-responsive smart device towards developing functionally cooperating systems. *Advances in Materials, 22*(45), 5125–5128.

[11]   Ju, G., Cheng, M., Xiao, M., Xu, J., Pan, K., Wang, X. *et al.* (2013). Smart Transportation Between Three Phases Through a Stimulus-Responsive Functionally Cooperating Device. *Advances in Materials, 25*(21), 2915–2919.

[12]   Dry, C. (1994). Matrix cracking repair and filling using active and passive modes for smart timed release of chemicals from fibers into cement matrices. *Smart Materials &Structures, 3*(2), 118.

[13]   Cho, S. H., White, S. R. & Braun, P. V. (2009). Self-Healing Polymer Coatings. *Advances in Materials, 21*(6), 645–649.

[14]   Burnworth, M., Tang, L., Kumpfer, J. R., Duncan, A. J., Beyer, F. L., Fiore, G. L., Weder, C. *et al.* (2011). Optically healable supra-molecularpolymers. *Nature, 472*(7343), 334–337.

[15]   Cordier, P., Tournilhac, F., Soulié-Ziakovic, C. & Leibler, L. (2008). Self-healing and thermoreversible rubber from supramolecular assembly. *Nature. 451*, 977 980.

[16]   Ying, H., Zhang, Y. & Cheng, J. (2014). Dynamic urea bond for the design of reversible and self-healing polymers. *Nature Communications. 5*, 3218.

[17]   Wang, G., Fu, Y., Guo, A., Mel, T., Wang, J., Ll, J. *et al.* (2017). Reduced graphene oxidepolyurethane nanocomposite foams as a reusable photo-receiver for efficient solarsteam generation. *Chemistry of Materials, 29*(13), 5629–5635.

[18]   Robinson, J. T., Tabakman, S. M., Liang, Y., Wang, H., Sanchez Casalongue, H., Vinh, D. *et al.* Ultrasmall Reduced Graphene Oxide with High Near-Infrared Absorbance for Photothermal Therapy. *Journal of the American Chemical Society, 133*(17), 2011)6825–6831.

[19]   Song, W. L., Cao, M. S., Lu, M. M., Bi, S., Wang, C. Y., Liu, J. *et al.* (2014). Flexible graphene/polymer composite films in sandwich structures for effective electromagnetic interference shielding. *Carbon, 66*(1), 67–76.

[20]   Shen, X., Gong, R. Z., Nie, Y. & Nie, J. H. (2005). Preparation and electromagnetic performance of coating of multiwall carbon nanotubes with iron nanogranule. *Journal of Magnetism and Magnetic Materials, 288*(288), 397–402.

[21]   Boland, C. S., Khan, U., Ryan, G., Barwich, S., Charifou, R., Harvey, A. *et al.* (2016). Sensitive electromechanical sensors using visco-elastic graphene-polymer nanocomposites. *Science, 354*(6317), 1257–1260.

[22]   Wang, Y., Wang, L., Yang, T., Li, X., Zang, X., Zhu, M. *et al.* (2014). Wearable and Highly Sensitive Graphene Strain Sensors for Human Motion Monitoring. *Advanced Functional Materials, 24*(29), 4666–4670.

[23]   Novoselov, K. S., Geim, A. K., Morozov, S. V., Jiang, D., Zhang, Y., Dubonos, S. V. *et al.* (2004). Electric Field Effect in Atomically Thin Carbon Films. *Science, 306*(5696), 666–669.

[24] Berger, C., Song, Z., Li, T., Li, X., Ogbazghi, A. Y., Feng, R. *et al.* (2004). Ultrathin Epitaxial Graphite: 2D Electron Gas Properties and a Route toward Graphene-basedNanoelectronics. *The Journal of Physical Chemistry B, 108*(52), 19912–19916.

[25] Berger, C. (2006). Electronic Confinement and Coherence in Patterned Epitaxial Graphene. *Science, 312*(5777), 1191–1196.

[26] Li, X., Cai, W., An, J., Kim, S., Nah, J., Yang, D. *et al.* (2009). Large-Area Synthesis of High-Quality and Uniform Graphene Films on Copper Foils. *Science, 324*(5932), 1312–1314.

[27] Zhang, L., Liang, J., Huang, Y., Ma, Y., Wang, Y. & Chen, Y. (2009). Size-controlled synthesis of graphene oxide sheets on a large scale using chemical exfoliation. *Carbon, 47*(14), 3365–3368.

[28] Liu, Z., Liu, Q., Huang, Y., Ma, Y., Yin, S., Zhang, X. *et al.* (2008). Organic Photovoltaic Devices Based on a Novel Acceptor Material: Graphene. *Advances in Materials, 20*(20), 3924–3930.

[29] Becerril, H. A., Mao, J., Liu, Z., Stoltenberg, R. M., Bao, Z. & Chen, Y. (2008). Evaluation of Solution-Processed Reduced Graphene Oxide Films as Transparent Conductors. *ACS Nanotechnology, 2*(3), 463–470.

[30] Hummers, W. S. Jr & Offeman, R. E. (1958). Preparation of graphitic oxide. *Journal of the american chemical society, 80*(6), 1339-1339.

[31] Samadi, N., Sabzi, M. & Babaahmadi, M. (2018). Self-healing and tough hydrogels with physically cross-linked triple networks based on Agar/PVA/Graphene. *International Journal of Biological Macromolecules. 107*, 2291–2297.

[32] Jing, X., Mi, H.-Y., Napiwocki, B. N., Peng, X.-F. & Turng, L.-S. (2017). Mussel-inspired electroactive chitosan/graphene oxide composite hydrogel with rapid self-healing and recovery behavior for tissue engineering. *Carbon. 125*, 557–570.

[33] Han, L., Lu, X., Wang, M., Gan, D., Deng, W., Wang, K. *et al.* (2017). A Mussel-Inspired Conductive, Self-Adhesive, and Self-Healable Tough Hydrogel as Cell Stimulators and Implan*Table* Bioelectronics. *Small (Weinheim an der Bergstrasse, Germany), 13*(2). http://doi:10.1002/smll. 201601916.

[34] Yanagisawa, Y., Nan, Y., Okuro, K. & Aida, T. (2018). Mechanically robust, readily repairablepolymers via tailored noncovalent cross-linking. *Science, 359*(6371), 72–76.

[35] Kirkby, E. L., Rule, J. D., Michaud, V. L., Sottos, N. R., White, S. R. & Manson, J. A. E. (2008). Embedded shape-memory alloy wires for improved performance of self-healing polymers. *Advanced Functional Materials, 18*(15), 2253–2260.

[36] Jones, A. S., Rule, J. D., Moore, J. S., Sottos, N. R. & White, S. R. (2007). Life extension of self-healing polymers with rapidly growing fatigue cracks. *Journal of the Royal Society Interface, 4*(13), 395–403.

[37] Klerk, L. A., Dankers, P. Y. W., Popa, E. R., Bosman, A. W., Sanders, M. E., Reedquist, K. A. & Heeren, R. M. A. (*2010*). *Analytical Chemistry, 82*, 4337.

[38] Kirkby, E. L., Michaud, V. J., Manson, J. A. E., Sottos, N. R. & White, S. R. (2009). Performance of self-healing epoxy with microencapsulated healing agent and shape memory alloy wires. *Polymer, 50*(23), 5533–5538.

[39] Moll, J. L., White, S. R. & Sottos, N. R. (2010). A self-sealing fiber reinforced composite. *Journal of Compos Mater*, In press, 10.1177/0021998309356605.

[40] Kessler, M. R., Sottos, N. R. & White, S. R. (2003). Self-healing structural composite materials. *Composites, A, 34*(8), 743–753.

[41] Liu, C. & Qin, H. (2007). Review of progress in shape-memory polymers, *17*. 1543–1558.

[42] Lendlein, A. & Kelch, S. (2002). Shape-Memory Effects in Polymer Networks Containing Reversibly Associating Side-Groups. *41*, 2034.

[43] Ratna, D. & Karger, J. (2008). Recent advances in shape memory polymers and composites: a review. *43*, 252.

[44] Meng, Q. & Hu, J. (2009). A review of shape memory polymer composites and blends. *40*, 1662.

[45] Liu, Y., Du, H., Leu, L. & Leng, J. (2014). Shape memory polymers and their composites in aerospace applications: a review. *27*, 528–532.

[46] Lendlein, A. & Langer, R. (2002). Biodegradable, Elastic Shape-Memory Polymers for Potential Biomedical Applications. *296*, 5976–5981.

[47] Seranno, M. & Carbajal, L. (2011). Novel Biodegradable Shape-Memory Elastomers with Drug-Releasing Capabilities. *29*, 1274.

[48] Weng, J. & Dai, L. (2013). Graphene enhances the shape memory of poly (acrylamide co acrylic acid) grafted and graphene, Macromol. *Rapid commune, 34.* 659–664.

[49] Karak, N. & Thakur, S. (2015). Tuning of sunlight induced self cleaning and self healing attribute of s nanocomposite by judicious compositional variation of the TiO_2. *Journal of Materials Chemistry A, 3.* 12334–12342.

[50] Mohammad, M. & Samand, N. (2016). Graphene network enabled high speed electrical actuation of shape memory nano composites based on PVA. *Polymer Interface, 66.* 665–671.

[51] Liu, F., Li, F., Deng, G., Chen, Y., Zhang, B. & Zhang, J. (2012). *Macromolecules, 45*(3), 1636–1645.

[52] Chambon, F., Petrovic, Z. S., Macknight, W. J. & Winter, H. H. M. (1986). *19*, 2146–2149.

[53] Takahashi, M., Yokoyama, K., Masuda, T. & Takigawa. (1994). The Journal of Chemical Physics, 101, 798–804.

[54] Tung, C. Y. M. & Dynes, P. J. J. (1982). *Applied Polymer Science, 27*, 569–574, 5. Levrand, B.; Fieber, W.; Lehn, J. M.; Herrmann, A. Helv. Chim. Acta 2007, 90, 2281–2314.

[55] Martin, J. E., Adolf, D. & Wilcoxon,. (1988). *Journal of the Physical Review Letters, 61*, 2620–2623.

[56] Martin, J. E., Adolf, D. & Wilcoxon,. (1989). *Journal of Plasma Physics, Physical Review A, 39*, 1325–1332.

[57] Burattini, S., Colquhoun, H. M., Fox, J. D., Friedmann, D., Greenland, B. W., Harris, P. J. F., Hayes, W., Mackay, M. E. & Rowan, S. J. (2009). A self-repairing, supramolecular polymer system: healability as a consequence of donor-acceptor [small pi]-[small pi] stacking interactions. *Chemical Communication*, 6717–6719.

[58] Hansen, C. J., Wu, W., Toohey, K. S., Sottos, N. R., White, S. R. & Lewis, J. A. (2009). Self-Healing Materials with Interpenetrating Microvascular Networks. *Advances in Materials, 21.* 4143–4147.

[59] Smith, M. E. & Dukey, M. D. (2013). Self healing Stretchable wires for reconfigurable circuit wiring and 3D microfluidics. *Advances in Materials, 25.* 1589–1592.

[60] Ge, L., Lu, C. & Ren, J. (2015). Graphene improved electrochemical property in self healing multilayer polyelectrolyte film, Colloids surf. A physiochemical. *Engineering Aspects, 465.* 26–31.

[61] Liu, N., Zou, Z. & Goo, Y. (2017). Highly stretchable and self healable super capacitors with reduced graphene oxide based fiber springs. *ACS Nanotechnology, 11.* 2066–2074.

[62] Chung, H. J. & Ok, L. (2017). Flexible and self healing aqueous super capacitor for low temperature application. *Scientific Reports, 7.* 1685.

[63] Lu, X. & Kong, J. (2017). Aqueous only green route self healable, UV resistant polyurethane, graphene/ lignin nano composite. *ACS Sustainable Chemistry & Engineering, 5.* 3148–3157.

[64] Goo, S. & Yang, J. (2017). Self assembled graphene oxide microcapsule in Pickering emulsion for self healing waterbone PU coatings. *Composite Science Technology, 151.* 282–290.

[65] Luckachan, G. E. & Mettol, V. (2016). Self healing protective coatings of polyvinyl butyl/ polypyrolle carbon black composite on carbon steel. *RSC Advances, 6.* 43237–43249.

[66] Patel, A. J., Sottos, N. R., Wetzel, E. D. & White, S. R. (2010). Autonomic healing of low-velocity impact damage in fiber-reinforced composites. *Composites, A, 41*(3), 360–368.

[67] Kessler, M. R. & White, S. R. (2001). Self-activated healing of delamination damage in woven composites. *Composites, A, 32*(5), 683–699.

[68] Sanada, K., Yasuda, I. & Shindo, Y. (2006). Transverse tensile strength of unidirectional fiber-reinforced polymers and self-healing of interfacial debonding. *Plast. Rubber Compos., 35*(2), 67–72.

[69]   Wilson, G. O., Moore, J. S., White, S. R., Sottos, N. R. & Andersson, H. M. (2008). Autonomic healing of epoxy vinyl esters via ring opening metathesis polymerization. *Advanced Functional Materials*, *18*(1), 44–52.

[70]   Karak, N. (2014). Green one step approach to prepare sulfur/ reduced graphene oxide nanohybrid for effective mercury ions removal. *RSC advances*, *5*. 2167–2176.

[71]   Hu, X. & Zhang, X. (2014). Double network self healing graphene hydrogel by two step method for anti cancer drug delivery. *Mater Tech: Adv Perform Mater*, *29*. 210–213.

[72]   Shi, G. & Bai, H. (2010). Three dimensional self assembly of graphene oxide and DNA into multi functional hydrogels. *ACS Nanotechnology*, *4*. 7358–7362.

[73]   Liu, K., Chan, C. W. & Tang, W. (2017). A mussel Inspired conductive; Self Adhesive and self healable tough hydrogel as cell stimulation and implantable bioelectronics. *Small*, *13*.

[74]   Yakacki, C. M., Shandas, R., Lanning, C., Rech, B., Eckstein, A. & Gall, K. (2007). *Biomaterials*, *28*, 2255–2263, PubMed: 17296222.

[75]   Feng, Y., Behl, M., Kelch, S. & Lendlein, A. (2009). *Macromolecular Bioscience*, *9*, 45–54, PubMed: 19089872.

[76]   Pinnavaia, T. J. (1983). Intercalated Clay Catalyst. *Science*, *220*, 365–371.

[77]   Peng, R. & Yang, Y. (2014). Conductive nanocomposite hydrogels with self healing property. *RSC Advances*, *4*. 35149–35155.

[78]   Peng, X. & Turung, L. S. (2017). Mussul Inspired electoactive chitosan/ graphene oxide hydrol with rapid self healing behavior for tissue engineering. Carbon, Vol. 125, pp. 557–570.

[79]   Yuan, C., Rong, M. Z., Zhang, M. Q., Zhang, Z. P. & Yuan, Y. C. (2011). Self-Healing of Polymers via Synchronous Covalent Bond Fission/Radical Recombination. *Chemistry of Materials*, *23*. 5076–5081.

[80]   Prager, S. & Tirrell, M. (1981). The healing process at polymer–polymer interfaces. *Journal of Chemical Physics*, *75*. 5194–5198.

[81]   Tong, Z. & liu, W. (2014). Fast self-healing of graphene oxide heatorite clay poly (N,N – dimethyl acrylamide) hybrid hydrogels realized by near infrared radiation. *ACS App Mater Interfaces*, *6*, 22855–22861.

[82]   Doddi, S., Ramakrishna, B., Venkatesh, Y. & Bangal, P. R. (2015). Photo-driven near-IR fluorescence switch: synthesis and spectroscopic investigation of squarine-spiropyran dyad. *RSC Advances*, *5*. 97681–97689.

[83]   Grady, B. P. (2011). Carbon Nanotube-Polymer Composites Manufacture, Properties, and Applications. John Wiley and Sons: New York, NY, USA, pp. 145.

[84]   Hou, C., Huang, T., Wang, H., Yu, H., Zhang, Q. & Li, Y. (2013). A strong and stretchable self-healing film with self-activated pressure sensitivity for potential artificial skin applications. *Scientific Reports*, *3*. 3138.

[85]   Faucheu, J., Gauthier, C., Chazeau, L., Cavaillé, J. Y., Mellon, V. & Lami, E. B. (2010). Miniemulsion polymerization for synthesis of structured clay/polymer nanocomposites: Short review and recent advances. *Polymer*, *51*, 6–17.

[86]   Tsardaka, E.-C., Stefanidou, M. & Pavlidou, E. (2018). The role of nanoparticles on the durability of lime-pozzolan binding system.

[87]   Fan, F., Zhou, C., Wang, X. & Szpunar, J. (2015). Layer-by-layer assembly of a self-healing anticorrosion coating on magnesium alloys. *ACS Applied Materials & Interfaces*, *7*. 27271–27278.

[88]   Lin, C., Sheng, D., Liu, X., Xu, S., Ji, F., Dong, L. *et al*. (2018). NIR induced self-healing electrical conductivity polyurethane/graphene nanocomposites based on Diels–Alder reaction. *Polymer*, *140*, 150–157.

[89]   Chipara, M. D., Chipara, M., Shansky, E. & Zaleski, J. M. (2009). Self-healing of high elasticity block copolymers. *Polymers for Advanced Technologies*, *20*(4), 427–431.

[90] Maiti, S., Shankar, C., Geubelle, P. H. & Kieffer, J. (2006). Continuum and molecular-level modeling of fatigue crack retardation in self-healing polymers. *Journal of Engineering Materials and Technology, Transactions of the ASME, 128*(4), 595–602.

[91] Yin, T., Rong, M. Z., Wu, J. S., Chen, H. B. & Zhang, M. Q. (2008). Healing of impact damage in woven glass fabric reinforced epoxy composites. *Composites, A, 39*(9), 1479–1487.

[92] Li, J., Zhang, G., Sun, R., & Wong, C. P. (2017). A covalently cross-linked reduced functionalized graphene oxide/polyurethane composite based on Diels–Alder chemistry and its potential application in healable flexible electronics. Journal of Materials Chemistry C, 5(1), 220–228.

[93] Keller, M. W., White, S. R. & Sottos, N. R. (2008). Torsion fatigue response of self-healing poly (dimethylsiloxane) elastomers. *Polymer, 49*(13–14).

[94] Cho, S. H., White, S. R. & Braun, P. V. (2009). Self-healing polymer coatings. *Advances in Materials, 21*(6), 645–649.

[95] Maiti, S. & Geubelle, P. H. (2006). Cohesive modeling of fatigue crack retardation in polymers: crack closure effect. *Engineering Fracture Mechanics, 73*(1), 22–41.

[96] Wilson, G. O., Caruso, M. M., Reimer, N. T., White, S. R., Sottos, N. R. & Moore, J. S. (2008). Evaluation of ruthenium catalysts for ring-opening metathesis polymerization-based self-healing applications. *Chemistry of Materials, 20*(10), 3288–3297.

[97] Wilson, G. O., Porter, K. A., Weissman, H., White, S. R., Sottos, N. R. & Moore, J. S. (2009). Stability of second generation Grubbs' alkylidenes to primary amines: formation of novel ruthenium-amine complexes. *Advanced Synthesis & Catalysis, 351*(11–12).

[98] Li, J., Feng, Q., Cui, J., Yuan, Q., Qiu, H., Gao, S., & Yang, J. (2017). Self-assembled graphene oxide microcapsules in Pickering emulsions for self-healing waterborne polyurethane coatings. Composites Science and Technology, 151, 282–290.

[99] Kamphaus, J. M., Rule, J. D., Moore, J. S., Sottos, N. R. & White, S. R. (2008). A new self-healing epoxy with tungsten (VI) chloride catalyst. *Journal of the Royal Society Interface, 5*(18), 95–103.

[100] Rong, M. Z., Zhang, M. Q., . & Zhang, W. (2007). A novel self-healing epoxy system with microencapsulated epoxy and imidazole curing agent. *Advanced Composites Letters, 16*(5), 167–172.

[101] Yin, T., Zhou, L., Rong, M. Z. & Zhang, M. Q. (2008). Self-healing woven glass fabric/epoxy composites with the healant consisting of microencapsulated epoxy and latent curing agent. *Smart Materials & Structures, 17*(1), 015019/1–8.

[102] Yin, T., Rong, M. Z., Zhang, M. Q. & Zhao, J. Q. (2009). Durability of self-healing woven glass fabric/epoxy composites. *Smart Materials & Structures, 18*(7), 074001.

[103] Thakur, S. & Karak, N. (2015). Tuning of sunlight-induced self-cleaning and self-healingattributes of an elastomeric nanocomposite by judicious compositional variation of theTiO2-reduced graphene oxide nanohybrid. *Journal of Materials Chemistry A, 3*(23), 12334–12342.

[104] Beiermann, B. A., Keller, M. W. & Sottos, N. R. (2009). Self-healing flexible laminates for resealing of puncture damage. *Smart Materials & Structures, 18*(8), 085001.

[105] Caruso, M. M., Delafuente, D. A., Ho, V., Sottos, N. R., Moore, J. S. & White, S. R. (2007). Solvent-promoted selfhealing epoxy materials. *Macromolecules, 40*(25), 8830–8832.

[106] Caruso, M. M., Blaiszik, B. J., White, S. R., Sottos, N. R. & Moore, J. S. (2008). Full recovery of fracture toughness using a nontoxic solvent-based self-healing system. *Advanced Functional Materials, 18*(13), 1898–1904.

[107] Zako, M. & Takano, N. (1999). Intelligent material systems using epoxy particles to repair microcracks and delamination damage in GFRP. *Journal of Intelligent Materials Systems and Structures, 10*(10), 836–841.

[108] Xiao, D. S., Yuan, Y. C., Rong, M. Z. & Zhang, M. Q. (2009). Self-healing epoxy based on cationic chain polymerization. *Polymer, 50*(13), 2967–2975.

[109] Lee, J. Y., Buxton, G. A. & Balazs, A. C. (2004). Using nanoparticles to create self-healing composites. *Journal of Chemical Physics, 121*, 5531–5540.
[110] Trau, M., Sankaran, S., Saville, D. A. & Aksay, I. A. (1995). Electric-field-induced pattern formation in colloidal dispersions. *Nature, 374*, 437–439.
[111] Ratna, D. & Karger-Kocsis, J. (2008). Recent advances in shape memory polymers and composites: A review. *Journal of Materials Science, 43*, 254–269.
[112] Yang, Y., Zhu, B., Yin, D., Wei, J., Wang, Z., Xiong, R., Shi, J., Liu, Z. & Lei, Q. (2015). Flexible self-healing nanocomposites for recoverable motion sensor. *Nano Energy. 17*, 1–9.

Index

https://doi.org/10.1515/9783111583716-007

www.ingramcontent.com/pod-product-compliance
Lightning Source LLC
Chambersburg PA
CBHW081523220326
41598CB00036B/6312